1286

Springer

Tokyo
Berlin
Heidelberg
New York
Barcelona
Budapest
Hong Kong
London
Milan
Paris
Santa Clara
Singapore

H.H. Natsuyama, S. Ueno, A.P. Wang

Terrestrial RadiativeTransfer

Modeling, Computation, and Data Analysis

With 72 Figures

 Springer

Harriet H. Natsuyama has published, as H. H. Kagiwada, six previous books and 150 papers on mathematical modeling, optimization, and radiative transfer, based on her work at the Rand Corporation, Hughes Aircraft Company, as Rockwell International Professor of Systems Engineering, and Foreign Special Visiting Professor in Japan.

Sueo Ueno is Professor Emeritus of Kyoto University and Member of the New York Academy of Sciences. The author of a book and several hundred papers on radiative transfer, he is Director of the Information Science Laboratory, Kyoto School of Computer Science.

Alan P. Wang is Professor of Mathematics at Arizona State University, where he conducts research in radiative transfer, approximations, inverse problems, integral equations, transport theory, and dynamics.

ISBN 4-431-70206-7 Springer-Verlag Tokyo Berlin Heidelberg New York

Printed on acid-free paper

© Springer-Verlag Tokyo 1998
Printed in Hong Kong

Typesetting: Camera-ready by authors
Printing & binding: Permanent Typesetting & Printing Co., Ltd., Hong Kong
SPIN: 10629979

To Our Families

Preface

In this book we share our work with those who are faced with the challenging problem of studying the earth's atmosphere and the interactions between the atmosphere and the earth's surface. While there are some excellent books on this topic written from the physical point of view, those discussing the modeling and computational aspects are few and far between. Our book is intended to bridge this gap so that students as well as investigators will be able to understand and apply practical ways of determining solutions.

Radiative transfer theory, on which this book is based, is elegant, and great minds have contributed to its richness. Instead of duplicating the classical references, we have taken a different approach: We have developed the invariant imbedding approach, both analytically and computationally, because of its attractiveness for producing numerical solutions. Having witnessed the transition to the computer age, we know that a new attitude to mathematical formulation is required. The one that we endorse is a model stated in the form of a Cauchy problem: a system of ordinary differential equations with a complete set of initial conditions. We chose this approach because it is well suited to implementation on digital computers.

This book gives a complete picture of the physical processes involved and the modeling of these processes into a mathematical form suitable for numerical solution. Computational methods such as those for numerical integration of systems of differential equations and for Gaussian quadrature are described and applied. Numerical results are presented in graphs and tables, and the behavior of solutions analyzed empirically.

Topics covered include internal and external radiation fields of inhomogeneous media with internal or with external illumination. Cases include isotropic and anisotropic scattering, omnidirectional and pencil illumination, flat and spherical geometry, and a variety of reflecting surfaces featuring rugged terrain. Some intriguing and unexpected relationships are discovered, developed, and tested, and these may be used to verify computational results as well as to simplify the determination of results for one case in terms of another.

We present a variety of methods for attacking and solving inverse problems of parameter estimation and remote sensing. Procedures are described for associative memories, quasilinearization, Monte Carlo methods, and simulation, and these procedures are applied to simulated and actual data.

Of special interest for terrestrial atmospheres are the treatments of atmospheric correction of spacecraft and aircraft data, remote sensing of rugged terrain and ground albedo, and atmospheric warming due to the greenhouse effect.

Advice is given on carrying out the types of computational studies discussed here. Hints are given on appropriate procedures for checking the computer program and executing the program with sensible choices of step size, quadrature formula, and other parameters. The computer procedures are described and sample computer programs are given, although the reader is welcome to choose other computer languages or software packages. These helpful hints are given for such topics as solving transfer problems via numerical integration of Cauchy problems and for conducting Monte Carlo studies of the estimation of unknown parameters.

New material is presented by the authors for the first time in book form. The use of associative memories for inverse problems and remote sensing is discussed, and recently obtained numerical results are given. We also present our work on the integral equations of model rendering. From our vast store of results we have collected and summarized major points and conclusions so that the reader has them readily at hand. The methodologies are described in sufficient detail that they can be implemented by a moderately capable programmer/analyst.

The overall plan of the book is as follows. The first half is devoted to basic concepts of radiative transfer in finite layered atmospheres and is suitable for inclusion in an undergraduate course on radiative transfer. Local scattering models are defined and multiple scattering models are described; multiple scattering functions of interest are introduced, and initial valued problems for these functions are derived using invariant imbedding. Effective and accurate computational methods for solving the integro-differential equations are described and abundant computational results are presented (Chapters 1 and 2).

Inverse problems in which unknown physical parameters are to be estimated on the basis of indirect, noisy measurements are considered in Chapter 3 and results of many experiments are provided. Direct and inverse problems of anisotropic scattering are taken up in Chapter 4. Chapter 5 complements Chapter 2 by presenting invariant imbedding derivations for reflection and transmission functions with one, two, ..., i..., scatterings.

The second half of the book begins with an overview of scattering matrix operator theory (Chapter 6) which is applied to remote sensing and inverse problems in Chapter 7 and Appendix D. Chapter 7 discusses the "de-blurring" of remotely sensed images by removal of effects of atmospheric scattering so that ground features can be better recognized. Determination of the terrain below an atmosphere is treated via "model rendering" in Chapter 8.

The "searchlight problem" in which a single narrow beam is incident on a scattering medium is discussed in Chapter 9. This is important not only for terrestrial atmospheres but in laser diagnosis and treatment in medicine. The mathematical formulation of the problem is a partial-differential-integral equation with six variables—a formidable computational problem. The final chapter models the spherical geometry of a planetary atmosphere. The partial differential-integral equation is solved as an initial value problem via several effective computational techniques.

The appendices serve as a reservoir of additional concepts, models, theory, and analysis. In Appendix A, a more rigorous basis of radiative transfer is presented. Then, an analytical method of deriving invariant imbedding equations is described in Appendix B. Appendix C shows how to model and solve the integral equation of multiple scattering describing the greenhouse effect and to determine temperatures in the atmosphere. Appendix D is an extension of scattering matrix theory to inverse problems.

The reader may select and dwell on various topics of interest in this book. We suggest that at least the first three chapters be carefully studied and that the basic physical models, mathematical formulations, and computational methods be understood before moving ahead.

We hope that the readers will find this book of value now and in the years ahead. We also hope that readers will share our fascination with these topics, and we wish fellow researchers good luck in their own investigations.

We owe the expert word processing work in this book to Linda Arneson and the excellent figures to Bruce Long. We are grateful to them not only for their skill but for their seemingly infinite patience in making the countless revisions that were requested. We offer our deepest appreciation to Professor Robert E. Kalaba for reviewing the manuscript and for the pleasure of our association with him over many years. We are also grateful to our families, friends, and colleagues who have supported our effort in many ways. H. H. N. would like to thank the Japanese Ministry of Education for the opportunity to spend several months in Japan in 1995, during which time this book was conceived and begun. S. U. would like to thank the Kyoto School of Computer Science and the Astrophysics Institute of Kyoto University for their kind assistance with this project. A. P. W. thanks the Department of Mathematics of Arizona State University for generously providing time and office support.

Sueo Ueno, Kyoto, Japan
Harriet H. Natsuyama, Yorba Linda, California, USA
Alan P. Wang, Tempe, Arizona, USA

Table of Contents

List of Figures

List of Tables

1. Basic Concepts

We introduce basic concepts for the modeling of radiative transfer using the invariant imbedding approach. We show, for a one-dimensional reflection problem, how an initial value problem is formulated. We obtain a differential equation with the independent variable being the thickness, and an initial condition, for thickness zero. We describe the numerical procedure for integrating this equation. Tables of reflection functions are presented. Cauchy-initial value-problems for source and internal intensity functions are also treated. This chapter serves as an introduction to the more advanced concepts in Appendix A, as well as the remaining chapters of this book.

1.1 Introduction

Classical radiative transfer theory was active until the late 1950's and culminated in the books by Busbridge, Chandrasekhar, and others [1–4]. After providing a framework for modelling the physical processes, these authors laid emphasis on the mathematical analysis of a linear two-point boundary value problem and its analytical solution for very thin or very thick homogeneous media and other limiting cases. This problem, when formulated this way, is very difficult to solve numerically due to its complexity and instability, and thus limits extension to more realistic models for data analysis. An alternative mathematical formulation is a Fredholm integral equation which, unfortunately, is ill-conditioned.

Classical radiative transfer has given way to modern radiative transfer in which the digital computer plays a driving role. John von Neumann during the fifties promoted the formulation of problems in mathematical physics as initial value problems for large systems of ordinary differential equations, for they would be amenable to solution by electronic computers, and this was found to be correct. Invariant imbedders answered this call to reconsider the formidable problems of radiative transfer [5, 6].

The invariant imbedding approach to mathematical physics comprises the following steps:

1. Modeling of physical processes
2. Derivation of exact initial value problem

3. Formulation of numerical method
4. Computational solution

Advantages of invariant imbedding became clear almost immediately. By the year 1965, the imbedders had computed many classical functions of radiative transfer for media of arbitrary thickness, stratification, and albedo for single scattering, with and without surface reflectors. The imbedding equations are stable and numerical techniques are highly advanced for producing precision results [7]. Further, the ability to solve direct problems leads to the solution of inverse problems of remote sensing, and the imbedding approach leads to analytical results of theoretical interest.

We refer to our earlier book on radiative transfer [7] as *Multiple Scattering Processes*.

The invariant imbedding approach used in this chapter has led to an analytical theory for integral equations [8]. Invariant imbedding is now used widely in many fields: control theory [9, 10]; differential equations and probability theory [11, 12], wave equations, quantum mechanics, and chemical physics [13–15]; inverse problems [16, 17]; transport theory [18], and in many other applications such as water resources [19] and geophysics [20], as well as in computing [21], and of course, radiative transfer [22, 23, 24].

We begin by applying basic modeling and invariant imbedding techniques to a simple problem in radiative transfer. We derive an initial value problem for reflection from a rod and we solve the problem computationally. Then, we present a description of the basic processes in radiative transfer that are used in modeling, and we introduce some of the important functions of radiative transfer. Detailed descriptions are provided in Appendix A.

1.2 Invariant imbedding and a simple model of reflection

We consider an inhomogeneous, one-dimensional medium - a rod - on which is incident at the right end a constant source of incident energy (particles of radiation or photons). We obtain a Cauchy system for reflected intensity. The initial condition depends on whether the left end of the rod is an absorber or a reflector of radiation.

$t = 0$ $\qquad\qquad\qquad\qquad\qquad\qquad\qquad t = x$ $\qquad$ Reflected particle, $r(x)$.
Incident particle.

Figure 1.1. A rod medium which is illuminated by a constant source at the right end.

Consider an inhomogeneous rod extending from $t = 0$ to x, as depicted in Fig. 1.1. Assume that when a particle moves through a distance Δ along the rod, there is a probability $\Delta + o(\Delta)$ that it will interact with the rod. The function $o(\Delta)$, of course, is such that the ratio $o(\Delta)/\Delta$ tends to zero as Δ tends to zero. The result of an interaction is that the particle is removed from the process. Then, with probability $\lambda(t)$, $0 \leq \lambda(t) \leq 1$, $0 \leq t \leq x$, it is re-emitted (scattered), and it is equally likely to go to the right or to the left (i.e., the scattering is isotropic). Alternatively, we say that λ is the fraction of scattered particles. In an inhomogeneous medium $\lambda(t)$ is a function of position, while in a homogeneous medium it is constant. Particles which emerge from the right or left end of the rod do not re-enter the rod. We say the rod is imbedded in free space. At the right end of the rod there is one incident particle per unit time. Let

$$r(x) \quad = \quad \text{the average rate at which particles are re-} \qquad (1.1)$$
flected from the rod of length x, due to one incident particle per unit time at the right end.

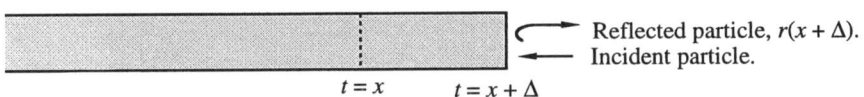

Reflected particle, $r(x + \Delta)$.
Incident particle.

$t = x$ $t = x + \Delta$

Figure 1.2. The rod of length $x + \Delta$ formed by addition of a segment of length Δ at the right end.

In the analysis to follow, x, the length of the rod is regarded as an independent variable, and $0 \leq x \leq L$, where L is a fixed maximum length. We conceptually increase the length of the rod from x to $x + \Delta$, as depicted in Fig. 1.2. We consider the rate at which particles are reflected from the slightly longer rod. In Fig. 1.3(a), the unit input of particles (to the left) is represented by the arrow labelled, "1." The fraction $1 - \Delta$ of these survive passage in going to the left through the rod of length Δ. The reflected fraction (to the right) is $(1 - \Delta)r(x)$, where $r(x)$ is the reflected rate for the rod of length x. This fraction is further reduced by the fraction $1 - \Delta$ in passing through the segment of length Δ going to the right. The overall result of this process is

$$\begin{aligned} 1 \cdot (1 - \Delta) \cdot r(x) \cdot (1 - \Delta) \\ = r(x) - \Delta r(x) - r(x)\Delta + r(x)\Delta^2 \\ = r(x) - 2\Delta r(x) + r(x)\Delta^2. \end{aligned} \qquad (1.2)$$

Figure 1.3(b) shows the rate of scattering in Δ. It represents processes in which unit incoming particles and the $r(x)$ reflected particles are scattered in the segment of length Δ. With probability, they interact, and the fraction $\lambda/2$ goes to the right (and an equal fraction to the left). This rate of production is equal to

$$[1 + r(x)]\Delta\frac{\lambda}{2}. \tag{1.3}$$

Of the scattered particles produced in Δ, the fraction (Fig. 1.3(c)), emerging from the right is unity when emerging directly, and equal to $r(x)$ when emerging after multiple scatterings in the rod, resulting in the terms

$$1 + r(x). \tag{1.4}$$

Combining the effects of production of scattering followed by emergence either directly or after multiple scatterings, we have from (1.3) and (1.4),

$$\frac{\lambda}{2}\Delta(1+r)(1+r) + o(\Delta), \tag{1.5}$$

(a)

(b)

(c)

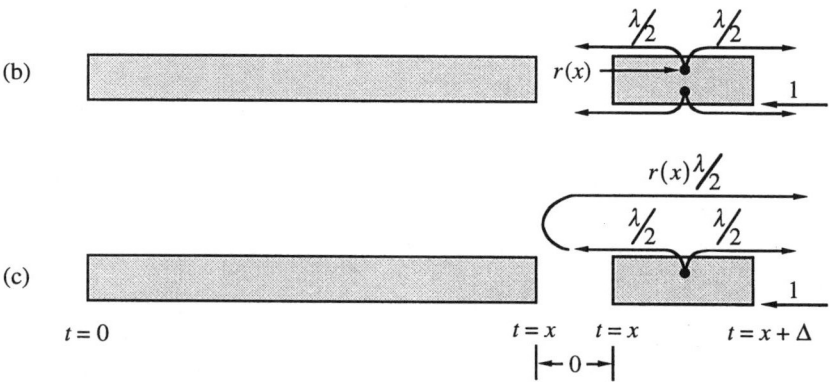

Figure 1.3. Absorption and scattering processes due to the added segment of the rod. (a) Absorption in Δ of incident and emergent particles. (b) Rate of production of scattered particles in Δ. (c) Emergence of particles after scattering in Δ.

where $r = r(x)$. Putting all the effects together and using (1.2) and (1.5), we have

$$r(x + \Delta) - r(x) = -2\Delta r(x) + \frac{\lambda}{2}\Delta(1+r)(1+r) + o(\Delta). \tag{1.6}$$

Divide through by Δ and let $\Delta \to 0$, using the definition of $o(\Delta)$. Then we obtain the differential equation

$$r'(x) = -2r + \frac{\lambda}{2}(1+r)^2, \tag{1.7}$$

a Riccati equation. The albedo for single scattering is evaluated at x, i.e., $\lambda = \lambda(x)$.

The initial condition which r satisfies is clearly

$$r(0) = 0, \tag{1.8}$$

which is due to the fact that no particles are reflected from a rod of length zero, since the rod is imbedded in free space. This is equivalent to having a perfect absorber at the left end. The initial condition depends upon the properties of scattering at the left end of the rod. If particles are scattered back into the rod, the initial condition will be different. Eqs. (1.7) and (1.8) form a Cauchy system, an initial value problem. This is readily resolved by modern analog and digital computers. The solution at $x = L$, namely $r(L)$, is the reflection function for the desired rod of length L. The solution for any rod of length $x \leq L$ is also obtained "along the way."

1.3 Computation of reflection function

To illustrate the basic concept for the numerical solution of a system of ordinary differential equations subject to initial conditions, we discuss the classical method of Euler in conjunction with the above equation for $r(x)$. We choose a step length h, sufficiently small for our purposes. We are given that $r(0) = 0$. The approximate value of r when $x = h$ is given by the formula

$$r(h) = r(0) + h\frac{dr}{dx}\bigg|_{x = 0}, \tag{1.9}$$

or

$$r(h) \cong r(0) + h\left[-2r(0) + \frac{\lambda}{2}(1+r(0))^2\right] = \frac{\lambda}{2}h, \tag{1.10}$$

where we have deleted the approximation sign. We may proceed to calculate $r(2h)$, ..., by use of the recurrence formula,

$$r((n+1)h) = r(nh) + h\left[-2r(nh) + \frac{\lambda}{2}(1+r(nh))^2\right], \tag{1.11}$$

for $n = 1, 2, \ldots$, until $n = N$ where $Nh = L$. Note that λ is evaluated at nh, i.e., $\lambda = \lambda(nh)$. Methods of great accuracy, stability and efficiency are available for the numerical integration of ordinary differential equations. The initial value problems of this book may be considered solvable by one of these methods. See, for example, [25].

We prepare a worksheet to carry out the Euler procedure for integrating the Riccati equation, starting with the zero initial condition. The following tables are from a printout for $\lambda = 2$. This corresponds to neutron multiplication in which two daughter neutrons are produced at each interaction. The first column is the current value of the length; the second is the corresponding Euler solution; the third is the analytical solution, $r(x) = \tan x$, and the fourth is the current derivative used in the Euler procedure. Table 1.1 shows the results using a step size of $h = 0.01$ for lengths $0.01, 0.02, \ldots, 0.20$ for 20 steps, while Table 1.2 is for step size 0.10. Observe that in Table 1.1, the agreement at $x = 0.20$ between the Euler solution and the theoretical is 0.2025 versus 0.2027, while that in Table 1.2 is 0.2010 versus 0.2027, a poorer agreement due to the larger step size. Observe that the derivative grows increasingly large, because the tangent function goes to infinity at $\pi/2$, and the numerical values become quite inaccurate.

Euler's method uses only the first term in the Taylor series expansion of the derivative function. Generally, accuracy improves with use of higher order terms; however, this is not practical, so methods using only first-order derivatives are preferred. Consider the differential equation,

$$dr/dx = f(x, r), \quad r(c) = \text{given}, \quad c \le x \le C. \tag{1.12}$$

The Runge-Kutta method is often used in the second-, third-, and fourth-order approximations [25]. In one of the popular forms, the second-order Runge-Kutta method produces the next step of the solution $r(x)$ by incrementing the current value $r(i)$ by using the average of the derivatives evaluated at $x(i)$ and at $x(i + 1)$,

$$r(i + 1) = r(i) + 0.5h[k(1) + k(2)], \tag{1.13}$$

where the first k, $k(1) = f(x(i), r(i))$, is the derivative evaluated at the current point, $(x(i), r(i))$, and the second k is the derivative evaluated at the point ahead, $(x(i) + h, r(i) + hk(1))$. The computational procedure remains the same: From the current point, compute the derivatives $k(1)$ and $k(2)$, then increment the solution according to the Runge-Kutta formula (1.13). This takes us to a new point, at which the procedure is repeated until the end of the desired interval is reached.

Let's compare these two algorithms for computing the reflection function when $\lambda = 1.0$, fixing the step size at 0.01. Table 1.3 shows an improvement of the second-order Runge-Kutta method over Euler's method.

Table 1.1. Euler's Method for Reflection Function. $\lambda = 2$, step size 0.01

x	Euler sol. $r(x)$	Anal. sol. $r(x)$	Deriv. $r'(x)$
0.00	0.0000	0.0000	1.0000
0.01	0.0100	0.0100	1.0001
0.02	0.0200	0.0200	1.0004
0.03	0.0300	0.0300	1.0009
0.04	0.0400	0.0400	1.0016
0.05	0.0500	0.0500	1.0025
0.06	0.0601	0.0601	1.0036
0.07	0.0701	0.0701	1.0049
0.08	0.0801	0.0802	1.0064
0.09	0.0902	0.0902	1.0081
0.10	0.1003	0.1003	1.0101
0.11	0.1104	0.1104	1.0122
0.12	0.1205	0.1206	1.0145
0.13	0.1307	0.1307	1.0171
0.14	0.1408	0.1409	1.0198
0.15	0.1510	0.1511	1.0228
0.16	0.1613	0.1614	1.0260
0.17	0.1715	0.1717	1.0294
0.18	0.1818	0.1820	1.0331
0.19	0.1921	0.1923	1.0369
0.20	0.2025	0.2027	1.0410
0.21	0.2129	0.2131	1.0453
0.22	0.2234	0.2236	1.0499
0.23	0.2339	0.2341	1.0547
0.24	0.2444	0.2447	1.0597
0.25	0.2550	0.2553	1.0650
0.26	0.2657	0.2660	1.0706
0.27	0.2764	0.2768	1.0764
0.28	0.2871	0.2876	1.0824
0.29	0.2980	0.2984	1.0888
0.30	0.3088	0.3093	1.0954

Table 1.2. Euler's Method for Reflection Function. $\lambda = 2$, step size 0.10

x	Euler sol. $r(x)$	Anal. sol. $r(x)$	Deriv. $r'(x)$
0.00	0.0000	0.0000	1.0000
0.10	0.1000	0.1003	1.0100
0.20	0.2010	0.2027	1.0404
0.30	0.3050	0.3093	1.0930
0.40	0.4143	0.4228	1.1717
0.50	0.5315	0.5463	1.2825
0.60	0.6598	0.6841	1.4353
0.70	0.8033	0.8423	1.6453
0.80	0.9678	1.0296	1.9367
0.90	1.1615	1.2602	2.3491
1.00	1.3964	1.5574	2.9499
1.10	1.6914	1.9648	3.8608
1.20	2.0775	2.5722	5.3159
1.30	2.6090	3.6021	7.8071
1.40	3.3898	5.7979	12.4905
1.50	4.6388	14.1014	22.5186
1.60	6.8907	-34.2325	48.4813
1.70	11.7388	-7.6966	138.7996
1.80	25.6188	-4.2863	657.3209
1.90	91.3509	-2.9271	8345.9790
2.00	925.9488	-2.1840	* * * * **

Table 1.3. Comparison of Computed Reflection Function for $\lambda = 1.0$

x	Euler $r(x)$	Runge-Kutta $r(x)$	Analytical $r(x)$
0.00	0.0000	0.0000	0.0000
0.10	0.0478	0.0453	0.0454
0.20	0.0913	0.0826	0.0913
0.30	0.1310	0.1134	0.1139

1.4 Internal intensity and source functions

1.4.1 Internal intensity model

Let us define the internal intensity functions

$$m(t, x) \quad = \quad \text{the average number of particles passing the} \quad (1.14)$$
$$\text{point } t \text{ which are traveling to the right in a}$$
$$\text{unit of time, in a rod of length } x, \text{ due to one in-}$$
$$\text{cident particle at the right end, per unit time.}$$

$$n(t, x) \quad = \quad \text{the average number of particles passing the} \quad (1.15)$$
$$\text{point } t \text{ which are traveling to the left in a unit}$$
$$\text{of time, in a rod of length } x, \text{ due to one inci-}$$
$$\text{dent particle at the right end, per unit time,}$$
$$0 \le t \le x.$$

We perform the same type of invariant imbedding as before. We add a thin segment to the right hand side of the rod of length x. The average number of particles passing the point t and traveling to the right in the augmented rod is:

$$m(t, x + \Delta) \quad = \quad m(t, x) - \Delta m(t, x) \qquad (1.16)$$
$$+ (\lambda/2)\Delta[1 + r(x)]m(t, x)$$
$$+ o(\Delta).$$

The first term on the right hand side is the average number in the rod of length x. The next term is the reduction in the number of particles at t due to the interaction of the incoming particles in the added segment. Particles that were on their way out at the right may interact in the small segment and be scattered back into the rod, causing a contribution to the particles at t; these give rise to the third term. All other processes account for terms of order higher than the first in Δ and are included in $o(\Delta)$. In the third term on the right, we have the same coefficient of $m(t, x)$ that we had before.

Consideration of the particles at the point t traveling to the left leads to an analogous equation,

$$n(t, x + \Delta) \quad = \quad n(t, x) - \Delta n(t, x) \qquad (1.17)$$
$$+ (\lambda/2)\Delta[1 + r(x)]n(t, x)$$
$$+ o(\Delta).$$

We carry out the limiting process as the thickness of the added segment tends to zero. Then we have the ordinary differential equations (t is held fixed)

$$m_x(t, x) = -m(t, x) + \frac{\lambda}{2}[1 + r(x)]\, m(t, x), \tag{1.18}$$

$$n_x(t, x) = -n(t, x) + \frac{\lambda}{2}[1 + r(x)]\, n(t, x), \tag{1.19}$$

$$x \le t \le L.$$

Observe the same factor

$$\frac{\lambda}{2}[1 + r(x)], \tag{1.20}$$

from Eq. (1.3), the rate of production of scattered particles at the right when the rod length is increased, where $\lambda = \lambda(x)$. When this is multiplied by $m(t, x)$, or $n(t, x)$, we have the contribution to the intensity at t.

The initial conditions, when the length $x = t$, are

$$m(t, t) = r(t), \tag{1.21}$$

$$n(t, t) = 1, \tag{1.22}$$

the former through the definitions of m and r, the latter through the description of the source. To form the complete initial value problem, we must adjoin the equations for r from Eqs. (1.7) and (1.8),

$$r'(x) = -2r + \frac{\lambda}{2}[1 + r]^2, \tag{1.23}$$

$$r(0) = 0. \tag{1.24}$$

To solve this enlarged Cauchy problem, we produce the solution to Eqs. (1.23) and (1.24) from $x = 0$ to $x = t$. At $x = t$, we adjoin Eqs. (1.18) and (1.19), and we use the initial conditions in Eqs. (1.21) and (1.22). We continue the solution of the entire system out to $x = L$. Then the internal intensities are known for fixed t and all $x \le L$.

1.4.2 Source function

The source function is defined as follows:

$$J(t, x) \quad = \quad \text{the rate of production of scattered particles} \tag{1.25}$$
going to the right (or left) at the point t in a rod of length x, due to one incident particle at the right end, per unit time.

The invariant imbedding equation with the segment of length Δ added to the right is:

$$J(t, x + \Delta) \;=\; J(t, x) - \Delta J(t, x) \tag{1.26}$$
$$+ (\lambda/2)\Delta[1 + r(x)]J(t, x)$$
$$+ o(\Delta).$$

Not surprisingly, this equation is similar to the previous ones. In the limit, the differential equation,

$$J_x(t, x) = -J(t, x) + (\lambda/2)[1 + r(x)]J(t, x) \tag{1.27}$$

is obtained. The initial condition used for J is based on the rate of scattering to the right or the left at the right end of the rod,

$$J(x, x) = (\lambda/2)[1 + r(x)] \tag{1.28}$$

and it is used when evaluated at $x = t$, i.e.,

$$J(t, t) = (\lambda/2)[1 + r(t)]. \tag{1.29}$$

To summarize sections 1.1 through 1.4, we remark that we have introduced the physical concept of invariant imbedding. We have derived a Cauchy system for reflection: a differential equation and an initial condition. We have demonstrated that a simple numerical integration method (algorithm) may be used to produce a computational solution. However, we would recommend that a fourth-order Runge-Kutta method be used instead for higher accuracy.

We have also obtained internal intensity functions and described how to compute them at any internal point in the medium from the enlarged Cauchy system. The source function, if desired, may also be computed. If, instead of an external source of radiation, there are internally emitted sources, varying with location in the medium, then the internal and external intensity functions may be determined in a similar fashion, from an even more expanded Cauchy system. See the following sections and the references.

1.5 Physical/mathematical descriptions

The study of radiative transfer was largely developed by astrophysicists studying stellar spectra, line formation, limb darkening, and other observed physical effects of scattering in stellar atmospheres. In these models, the sources of radiation were likely to be due to black body radiation and the scattering medium could be regarded as being semi-infinitely thick. Newer models appropriate for terrestrial atmospheres account for illumination from the outside, finite atmospheric thickness, inhomogeneities of local scattering properties, and surface reflection by different types of land cover or water.

For the purposes of this book, let us regard the earth's atmosphere as a plane-parallel medium(a slab, or layer), of infinite horizontal extent, which is vertically stratified, i.e., having scattering properties that depend upon the height (altitude, or level) above the ground. The inhomogeneities may also extend to dependence on horizontal coordinates as well. The "ground" is modeled as a perfect absorber, a Lambert diffusely reflecting surface, a specular reflector, or other reflector, whose reflecting properties may vary from point to point on the surface. For example, we may have a boundary demarcating a land surface from ocean, or the terrain of the land may vary from smooth to rugged.

Fig. 1.4 is a sketch representing the radiation field in the medium as well as radiation emerging from the top and from the bottom of a slab.

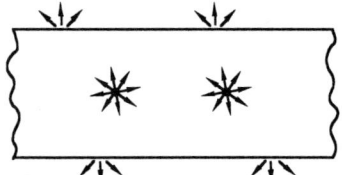

Figure 1.4. Radiation fields of a finite slab.

The atmosphere under consideration is an absorbing and scattering medium. A beam of radiation travelling in a medium is reduced in two ways: it is absorbed by the medium and turned into other forms of energy such as heat, and it is scattered into other directions. (While radiation can be scattered into other frequencies, such problems are outside the scope of this book.) The absorption coefficient is defined such that

$$\alpha dr = \text{the fraction of energy absorbed by the matter} \qquad (1.30)$$
$$\text{in the medium as a beam of radiation travels}$$
$$\text{through a cylinder of unit cross-sectional area}$$
$$\text{and height } dr, \text{ per unit mass of the matter,}$$

and similarly for the scattering coefficient σ. We may regard the scattering event as being the absorption of the fraction α of the beam followed by a scattering into all directions (isotropically or otherwise) with probability λ, where

$$\lambda = \sigma/\alpha \text{ the albedo for single scattering, i.e., frac-} \qquad (1.31)$$
$$\text{tion of interacting radiation that is scattered.}$$

In an inhomogeneous medium, these coefficients and the albedo for single scattering vary from point to point.

In scattering media, it is convenient to describe positions by optical, rather than geometrical, units. Radiation traversing an optical distance is reduced along the path: The optical distance, or optical height, is measured such that in a small optical distance interval dz the fraction of radiation absorbed is simply dz, to first order. The vertical optical distance between two geometrical points at levels $r = 0$ and $r = L$ is given by the integral

$$z = \int_0^L \alpha(r)\rho(r)dr, \tag{1.32}$$

assuming only a variation with altitude r, and letting ρ = density of the medium.

When radiation travels a small optical distance Δ, the probability of an interaction of the radiation with the matter in the medium is $\Delta + o(\Delta)$, where $o(\Delta)$ includes terms of order second and higher in Δ, such that as $\Delta \to 0$, $o(\Delta)/\Delta$ also $\to 0$:

$$\Delta \; = \; \text{optical distance, i.e., distance in which the} \tag{1.33}$$
$$\text{probability of an interaction of radiation with}$$
$$\text{the medium is } \Delta + o(\Delta).$$

If the vertical height is Δ and the direction of travel has direction cosine v, then the probability of an interaction is Δ/v and the fraction of radiation that survives is $1 - \Delta/v$. By integration, we know the change in radiative energy over a finite optical distance: When radiation travels a finite vertical optical distance z, only the fraction $\exp(-z)$ survives. If the radiation is travelling at an angle making the angle $\theta = \arccos v$ with the vertical, the fraction is $\exp(-z/v)$. This situation is sketched in Fig. 1.5.

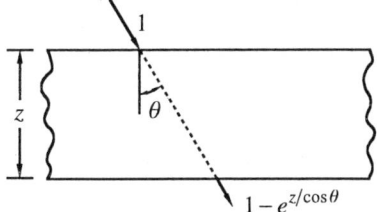

Figure 1.5. Reduction of radiation by absorption along a path.

For terrestrial and planetary atmospheres, it is often suitable to model the solar illumination as uniform parallel rays incident on the top of the layer at a constant angle from the vertical. For convenience, we frequently indicate the direction by u, the cosine of the polar angle shown in Fig. 1.6(a).

There may also be emitting sources of radiation in the medium, distributed according to position in the medium. These emissions may be due to the greenhouse effect and heating in the atmosphere, or other primary sources of radiation; these are sketched in Fig. 1.6(b).

When uniform parallel rays of radiation pass through a medium without scattering, the energy drops off exponentially, from optical level x to optical level $z \leq x$, according to the exponential law,

$$\exp[-(x - z)/u].$$

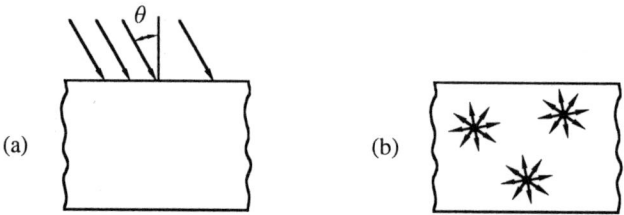

Figure 1.6. (a) Monodirectional illumination of top layer. (b) Internal sources.

If the incident energy is π per unit normal area with direction cosine u, then the energy per unit horizontal area is πu. In a cylinder of unit horizontal area and height Δ, this reduced radiation causes this amount of scattering per unit time per unit solid angle per unit frequency interval:

$$(\pi u)\exp[-(x-z)/u](\Delta/u)(\lambda(z)/4\pi) = (\lambda(z)/4)\exp[-(x-z)/u]\Delta, \quad (1.34)$$

where

$$
\begin{aligned}
\Delta/u \;&=\; \text{slant optical distance travelled} \\
\lambda \;&=\; \text{albedo for single scattering} \\
&=\; \text{fraction of interacting radiation that is scattered} \\
4\pi \;&=\; \text{solid angle, in steradians.}
\end{aligned}
$$

The primary rate of scattering per unit volume per unit solid angle per unit frequency interval with this incident source is

$$(\pi u)\exp[-(x-z)/u](1/u)(\lambda(z)/4\pi) = (\lambda(z)/4)\exp[-(x-z)/u]. \quad (1.35)$$

1.6 Intensity of radiation and source function

In the study of radiative transfer one wants to determine the intensity of radiation in a medium. Intensity I has units of energy per unit normal area per unit solid angle per unit time per unit frequency interval. Consider a beam of radiation of a certain intensity travelling through a medium in a certain direction denoted by the angle θ measured with respect to the upward vertical and the azimuth angle ϕ. The beam extends over a small solid angle, $d\Omega$, where $d\Omega = \sin\theta d\theta d\phi$, and passes through a small area ds normal to the direction of propagation. This is depicted in Fig. 1.7. The amount of energy contained is

$$dE = I d\Omega ds dt df = I\sin\theta d\theta d\phi ds dt df \quad (1.36)$$

where dt is the time interval and df is the frequency interval. If we want the energy passing through a unit *horizontal* area, we must multiply the right hand side by $\cos\theta$. At a given point in the medium, the intensity depends at least on the polar angle so we write $I = I(\theta)$; it will also depend on the coordinates of the point and other variables. Historically, the intensity function is regarded as the solution of a two-point boundary value problem. A more modern view is to compute the intensity as the solution of an initial value problem, as was done in Section 1.4. For the layered slab, this is treated in Chapter 2.

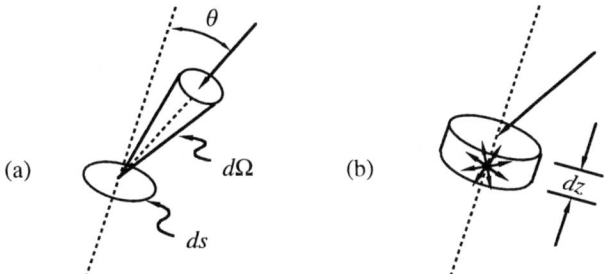

Figure 1.7. (a) Intensity of radiation. (b) Rate of production of scattered radiation.

An important quantity in the theory of radiative transfer is the source function which gives the rate at which energy is produced in a unit volume per unit solid angle per unit frequency interval. Consider the production of scattered radiation in a cylinder with unit horizontal cross-sectional area and optical height dz. The amount of scattered energy produced per unit time per unit frequency interval is made up of two parts: that due to the primary production or specular part, and that due to the scattering of radiation passing through the volume, or diffused part. For the sake of simplicity, we assume here that the scattering is isotropic, i.e., scattered radiation is uniformly distributed into all directions. If we let the primary production rate be denoted $g(z)$, and we note that the other term is an integral of the radiation that passes through, interacts and is scattered, namely,

$$\int_0^\pi \int_0^{2\pi} \{I(\theta)\cos\theta\sin\theta d\theta d\phi dz/\cos\theta\}\lambda(z)/4\pi \qquad (1.37)$$

$$= [\lambda(z)/4\pi] \int_0^\pi \int_0^{2\pi} I(\theta) sin\theta d\theta d\phi dz$$

$$= [\lambda(z)/4\pi] \int_{-1}^1 \int_0^{2\pi} I(\theta(v)) dv d\phi dz$$

$$= [\lambda(z)/2] \int_{-1}^1 I(\theta(v)) dv dz,$$

assuming that the intensity does not depend on the azimuth or the level. Therefore, the source functionis expressible as

$$J(z) = g(z) + [\lambda(z)/2] \int_{-1}^{1} I(\theta(v)) dv. \qquad (1.38)$$

We may see this equation written in the form,

$$J(z) = g(z) + [\lambda(z)/2] \int_{-1}^{1} I(v) dv. \qquad (1.39)$$

According to this formula, the source function can be determined from the intensity function by integrating over the direction cosine.

The intensity function can in turn be expressed as an integral of the source function over distance. The intensity of radiation at a point at altitude z propagating in a direction whose cosine is v, $I(z, v)$, is due to the contribution of all the scattered radiation along the path with direction cosine v. In other words, it is the integral of the radiation that was last scattered at the general altitude z' and has been attenuated until it reached z. The upwelling intensity is expressed as

$$I(z, +v) = \int_{0}^{z} J(z') dz' \exp(-z'/v)(1/v), \qquad (1.40)$$

and the downwelling as

$$I(z, -v) = \int_{z}^{x} J(z') dz' \exp(-(z' - z)/v)(1/v). \qquad (1.41)$$

Here $z = x$ is the top of the layer and $0 < v \le 1$. The factor $1/v$ converts the unit horizontal area to the unit normal area that the definition of intensity requires. While these expressions are useful for analytical purposes, they are not practical computationally because the integrand varies too sharply for customary quadrature methods to be effective. Still, they are useful for numerical approximations. In the classical theory of radiative transfer, the source function was determined as the solution of a Fredholm integral equation, whereas now we evaluate the source function through the initial value method of Chapter 2.

1.7 Discussion

We have discussed basic ideas in radiative transfer as well as in the modeling and computing aspects of invariant imbedding.

Books on radiative transfer include those previously referenced as well as [26–34].

The physical, modeling, mathematical, and computing concepts introduced in this chapter are expanded in the chapters that follow. In particular, we consider a radiative transfer process in an inhomogeneous plane-parallel medium.

References

1. I. W. Busbridge, *The Mathematics of Radiative Transfer*, Cambridge University Press, Cambridge, 1960.

2. S. Chandrasekhar, *Radiative Transfer*, Dover Publications, New York, 1960.

3. V. Kourganoff, *Basic Methods in Transfer Problems*, Dover Publications, New York, 1963.

4. V. V. Sobolev, *Light Scattering in Planetary Atmospheres*, Pergamon Press, New York, 1974.

5. R. Bellman, R. Kalaba and M. Prestrud, *Invariant Imbedding and Radiative Transfer in Slabs of Finite Thickness*, American Elsevier Publishing Co., New York, 1963.

6. R. E. Bellman, H. H. Kagiwada, R. E. Kalaba and M. C. Prestrud, *Invariant Imbedding and Time-Dependent Transport Processes*, American Elsevier Publishing Co., New York, 1964.

7. H. Kagiwada, R. Kalaba and S. Ueno, *Multiple Scattering Processes: Inverse and Direct*, Addison-Wesley Publishing Company, Reading, Mass., 1975.

8. H. H. Kagiwada and R. E. Kalaba, *Integral Equations via Invariant Imbedding*, Addison-Wesley Publishing Co., Reading, Mass., 1975.

9. A. P. Sage, "Invariant Imbedding in Control, Estimation, and System Identification," *Appl. Math. Comput.*, Vol. 45, 1991, p. 99.

10. J. L. Calvet and G. Viargues, "Invariant Imbedding and Parallelism in Dynamic Programming for Feedback Control," *J. Optimiz. Theory Appl.*, Vol. 87, 1995, p. 121.

11. J. Garnier, "Stochastic Invariant Imbedding. Application to Stochastic Differential Equations with Boundary Conditions," *Prob. Theory & Related Fields*, Vol. 103, 1995, p. 249.

12. I. S. Ayoubi, "The Riemann-Green Function and the Invariant Imbedding Equations for Hyperbolic Systems of First-order," *Appl. Math. Comp.*, Vol. 55, 1993, p. 101.

13. M. E. Davison and R. C. Winther, "A General Approach to Splitting and Invariant Imbedding for Linear Wave Equations," *J. Math. Anal. Appl.*, Vol. 188, 1994, p. 158.

14. A. J. Haines and M. V. deHoop, "An Invariant Imbedding Analysis of General Wave Scattering Problems," *J. Math. Phys.*, Vol. 37, 1996, p. 3854.

15. Y. B. Band and I. Tuvi, "Quantum Rearrangement Scattering Calculations Using the Invariant Imbedding Method," *J. Chem. Phys.*, Vol. 100, 1994, p. 8869.

16. J. Corones and Z. Sun, "Simultaneous Reconstruction of Material and Transient Source Parameters Using the Invariant Imbedding Method," *J. Math. Phys.*, Vol. 34, 1993, p. 1824.

17. S. He and S. Strom, "The Electromagnetic Inverse Problem in the Time Domain for a Dissipative Slab and a Point Source Using Invariant Imbedding:

Reconstruction of the Permittivity and Conductivity," *J. Comp. Appl. Math.*, Vol. 42, 1992, p. 137.

18.   S. K. Srinivasan and R. Vasudevan, "Particle Multiplicity Distribution à la Invariant Imbedding and Natural Scaling," *Comp. & Math. Appl.*, Vol. 22, 1991, p. 59.

19.   V. H. Weston, "Invariant Imbedding for the Wave Equation in Three Dimensions and the Applications to the Direct and Inverse Problems," *Inverse Problems*, Vol. 6, 1990, p. 1075.

20.   G. Nadimuthu and E. S. Lee, "Invariant Imbedding Filter in the Modeling of Water Resources," *Comp. & Math. Appl.*, Vol. 21, 1991, p. 9.

21.   D. E. Womble, R. C. Allen, Jr. and L. S. Baca, "Invariant Imbedding and the Method of Lines for Parallel Computers," *Parallel Computing*, Vol. 11, 1989, p. 263.

22.   M. I. Mischenko, "The Fast Invariant Imbedding Method for Polarized Light: Computational Aspects and Numerical Results for Rayleigh," *J. Quant. Spectrosc. Radiat. Transfer*, Vol. 43, 1990, p. 163.

23.   A. P. Wang, "Basic Equations of Three Dimensional Radiative Transfer," *Journal of Mathematical Physics*, Vol. 31, No. 10, 1990, p. 175.

24.   S. Ueno and A. P. Wang, "Invariant Imbedding and Order-of-Scattering Theory in Radiation Field," *Comp. & Math. Appl.*, Vol. 27, 1994, p. 175.

25.   B. Carnahan, H. Luther and J. Wilkes, *Applied Numerical Methods*, Wiley, New York, 1969.

26.   E. G. Yanovitskij, *Light Scattering in Inhomogeneous Atmospheres*, Springer-Verlag, New York, 1996.

27.   Jacqueline Lenoble, *Atmospheric Radiative Transfer*, A. Deepak Pub., 1993.

28.   K. Y. Kondratyev, V. V. Kozoderov and O. I. Smokty, *Remote Sensing of the Earth from Space: Atmospheric Correction*, Springer-Verlag, New York, 1992.

29.   G. A. D'Almeida, P. Koepke and E. P. Shettle, *Atmospheric Aerosols: Global Climatology and Radiative Characteristics*, A. Deepak Pub., 1991.

30.   W. Kalkofen, ed., *Numerical Radiative Transfer*, Cambridge University Press, Cambridge, 1988.

31.   Jacqueline Lenoble, ed., *Radiative Transfer in Scattering and Absorbing Atmospheres: Standard Computational Procedures*, A. Deepak Pub., 1985.

32.   H. C. Van de Hulst, *Light Scattering by Small Particles*, Dover, New York, 1982.

33.   R. Bellman and G. M. Wing, *An Introduction to Invariant Imbedding*, Soc. Indus. Appl. Math., 1962.

34.   C. F. Bohren, ed., *Selected Papers on Scattering in the Atmosphere*, SPIE Optical Engineering Press, Bellingham, WA, 1989.

2. Inhomogeneous Plane-Parallel Atmospheres

This chapter builds upon the physical descriptions of scattering processes of Chapter 1. It develops via invariant imbedding techniques effective mathematical and computational models of diffuse reflection and transmission due to multiple scattering in vertically stratified media. It treats the determination of internal diffuse intensities without use of the unstable equations of transfer and the computation of source functions without having to solve their ill-conditioned integral equations. Furthermore, internal and external intensity fields as well as source functions due to vertically inhomogeneous distributions of emitting sources are obtained. The emphasis here is to obtain exact Cauchy problems which are well solved computationally and to present samplings of the extensive numerical results that have been obtained. Cauchy problems are initial value problems for systems of differential equations and are attractive for computational solution.

2.1 Introduction

The previous chapter described the basic processes of scattering of radiation in media and defined many of the fundamental physical quantities in which we are interested. Some of the physical quantities are parameters which define the problem at hand. These are usually the single scattering parameters. Others are functions which are to be determined after modeling and solution of the model equations. These are functions which summarize the results of multiple scatterings. These functions have several independent variables, and they satisfy rather formidable functional equations.

These functional equations can be tamed, as we shall see in this chapter. The physical problem of this chapter is radiative transfer in inhomogeneous flat layers which are often referred to as plane-parallel media or slabs. Traditional model equations are the two-point boundary value problem and the Fredholm integral equation. The modern model equations which we present are initial value problems, also called Cauchy problems. The advantage of the modern over the traditional is the computability of solutions using this formulation.

Invariant imbedding is a technique for the modeling of physical problems that leads to Cauchy problems. One version of invariant imbedding is the physical one in which the Cauchy problem is derived directly from physical principles (such as those discussed in Chapter 1). An alternate derivation begins with a traditional model such as a boundary value problem or an integral equation which is then analytically manipulated to yield the corresponding Cauchy problem. The invariant imbedding Cauchy problem is an exact equivalence of the underlying boundary value problem or integral equation. The beauty of invariant imbedding is that the equations are stable and can be solved to a high degree of computational precision.

This chapter takes up the following cases: reflected and transmitted intensities for a slab illuminated by uniform parallel rays of radiation at the top; the computational method for the solving of Cauchy problems and the presentation of numerical results; slabs in which there is a variable distribution of emitting sources of radiation and the determination of internal intensities and source functions; the effect of reflecting surfaces below the slabs and some fascinating relationships with other cases; and finally the case of isotropic sources of radiation uniformly distributed at the upper surface of a slab and its importance and its relationship to other cases.

An attractive feature of invariant imbedding is the parametric study which automatically ensues—a study in how a function varies with increasing optical thickness. The optical thickness of the slab is increased from zero to any finite value, and the multiple scattering solutions for all intermediate thicknesses are obtained. Prior to the use of invariant imbedding in radiative transfer, the only known numerical results were those for very thin slabs or for very thick ones.

Another attractive feature of invariant imbedding is that this ability to solve direct problems (those problems in which physical parameters are given) enables us to attack the inverse problems of remote sensing: those in which we estimate the parameters based on multiple scattering observations.

2.2 Diffuse Reflection and Transmission

2.2.1 The physical problem

We consider radiative transfer in a plane-parallel, inhomogeneous medium which scatters radiation isotropically. The slab extends indefinitely in horizontal directions, while it has a finite vertical thickness. The albedo for single scattering varies with optical altitude. The slab is uniformly illuminated at the top by parallel rays of radiation of net flux π per unit normal area. This is the unit input of radiation. Since diffuse intensity is defined to be the intensity of radiation that has undergone one or more scatterings, it is clear that any diffuse radiation field – whether in the medium or external to it – is independent of the azimuth angle. Thus the incident direction is specified by only a polar angle, the angle whose cosine is z. Let the optical thickness be x, $0 \leq x \leq x_1$. The optical altitude is denoted t, $0 \leq t \leq x$. See Fig. 2.1. The albedo for single scattering is the function, $\lambda(t)$. The intensity of diffusively reflected radiation, i.e., the radiation emerging from the top of the slab in the direction whose cosine is v is $r(x, v, z)$. The surface at the bottom reflects diffusely. In the extreme case when it reflects nothing, it is a perfect absorber.

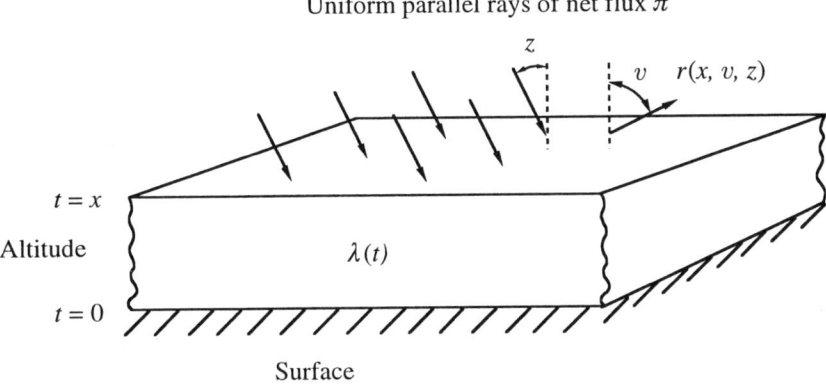

Figure 2.1. Monodirectional illumination of an inhomogeneous slab and diffusely reflected radiation, $r(x, v, z)$.

2.2.2 Invariant imbedding

We perform invariant imbedding by comparing two problems that are identical except that the optical thickness of one is slightly greater than the other by an amount Δ. We compare the intensities of reflected radiation from these two slabs, as shown in Fig. 2.2. The difference in the intensities of radiation reflected from the slab of thickness $x + \Delta$ and thickness x is due to the absorption and scattering processes that take place in the layer of thickness Δ. Consider Fig. 2.3(a), which depicts incoming radiation being absorbed in Δ, diffusely reflected in x, and absorbed in Δ. This amount of radiation is

$$[1 - \Delta z^{-1}]r(x, v, z)[1 - \Delta v^{-1}]. \tag{2.1}$$

This term is derived as follows. Incident radiation is reduced along its slant path in the thin layer by the factor:

$$1 - \Delta z^{-1}. \tag{2.2}$$

This radiation is reflected from the slab of thickness x in amount

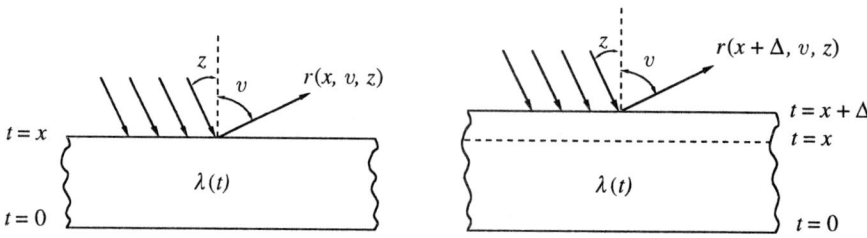

Figure 2.2. Side view of an inhomogeneous slab whose optical thickness is increased from x to $x + \Delta$, and a representative of the respective reflected intensities in direction arccos v.

$$[1 - \Delta z^{-1}]r(x, v, z). \tag{2.3}$$

However this radiation is reduced during its slant passage out through Δ, thus resulting in

$$[1 - \Delta z^{-1}]r(x, v, z)[1 - \Delta v^{-1}]. \tag{2.4}$$

This may be simplified to

$$\begin{aligned} r(x, v, z)[1 - \Delta z^{-1} &- \Delta v^{-1} + o(\Delta)] \\ = r(x, v, z)[1 &- \Delta(z^{-1} + v^{-1}], \end{aligned} \tag{2.5}$$

where we have omitted the higher order term in Δ.

Next consider, in Fig. 2.3(b), the rate at which radiation is scattered in a cylinder with unit horizontal area and height Δ. This rate is due to two types of scattering events. (1) An amount of incident radiation will be scattered at the rate

$$\pi\lambda(x)\Delta \tag{2.6}$$

into all directions, of which there are 4π steradians. Then the rate of scattering per unit steradian is

$$\frac{\pi\lambda(x)}{4\pi}\Delta = \frac{\lambda(x)}{4}\Delta. \tag{2.7}$$

(2) Emergent radiation from slab x going in a direction whose cosine is between v' and $v' + dv'$ will also be scattered with fraction $\lambda(x)$ into 4π

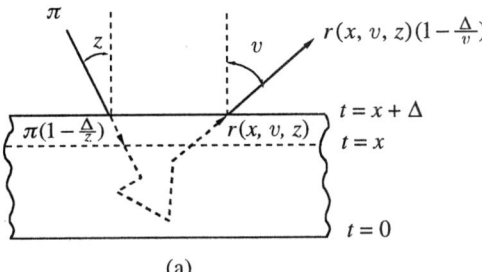

(a)

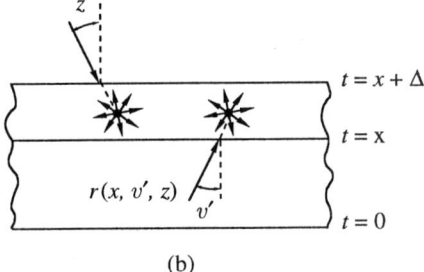

(b)

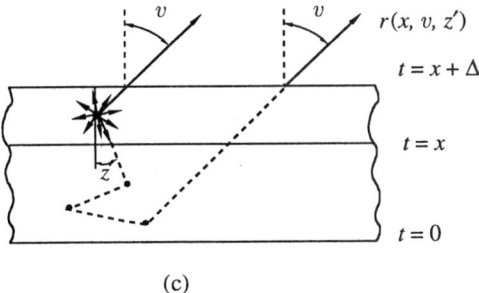

(c)

Figure 2.3. Production of scattered radiation at the top, by scattering of incident and emergent radiation. a) Reduction of reflected intensity due to absorption (at the top) of incident and emergent radiation. b) Production of scattered radiation at the top, by scattering of incident and emergent radiation. c) Reflection of radiation scattered into the direction $\arccos v$ either directly or after multiple scatterings, after scattering at the top.

steradians, and there will be 2π of such radiation. This accounts for the quantity, in units of π, of

$$\frac{r(x, v', z)2\pi dv'}{\pi}\Delta = 2r(x, v', z)dv'\Delta. \tag{2.8}$$

This is integrated over v' and multiplied by the fraction $\lambda(x)/4$ to give

$$\frac{\lambda(x)}{4}2\int_0^1 r(x, v', z)dv'\Delta. \tag{2.9}$$

Combining the terms from effects (1) and (2), we have the rate of production of scattered radiation at the top per unit solid angle in the cylinder of volume Δ:

$$\frac{\lambda(x)}{4}\Delta\left\{1 + 2\int_0^1 r(x, v', z)dv'\right\}. \tag{2.10}$$

This rate of production of scattered radiation at the top serves as a source of uniform *isotropic* illumination on the slab of thickness x, as shown in Fig. 2.3(b), and the reflection of this radiation, shown in Fig. 2.3(c). Radiation may emerge from the top directly with intensity v^{-1}. Or, radiation may enter the slab of thickness x with direction cosine z' and emerge after diffuse reflection. When writing this term, we must properly normalize the incident radiation by dividing it by $\pi z'$ and we must also integrate over the azimuth directions while the polar angle varies from $\arccos z'$ to $\arccos z' + dz'$. Thus the two terms are

$$v^{-1} + \int_0^1 (\pi z')^{-1}r(x, v, z')2\pi dz'. \tag{2.11}$$

The combined multiple scattering term of Figs. 2.3(b) and 2.3(c) is obtained from (2.10) and (2.11):

$$\frac{\lambda(x)}{4}\Delta\left\{1 + 2\int_0^1 r(x, v', z)dv'\right\}\left\{v^{-1} + 2\int_0^1 r(x, v, z)dz'/z'\right\}. \tag{2.12}$$

We can now express the functional relationship between r for thickness x and r for thickness $x + \Delta$. On the one hand the difference between the two intensities is given by $r(x + \Delta, v, z) - r(x, v, z)$. On the other hand, we have determined that the physical changes are given by the sum of Eqs. (2.5) and (2.12). Equating the two relationships, we have the integro-difference equation

$$r(x + \Delta, v, z) - r(x, v, z)$$
$$= r(x, v, z)[1 - \Delta(z^{-1} + v^{-1})] \tag{2.13}$$
$$+ \frac{\lambda(x)}{4}\Delta\left\{1 + 2\int_0^1 r(x, v', z)dv'\right\}\left\{v^{-1} + 2\int_0^1 r(x, v, z')dz'/z'\right\}$$
$$+ o(\Delta).$$

Division by Δ and passage to the limit as Δ tends to 0 leads to the integro-differential equation for $r(x,v,z)$,

$$
\begin{aligned}
r_x(x,v,z) &= -(z^{-1}+v^{-1})r(x,v,z) \\
&+ \frac{\lambda(x)}{4}\left\{1+2\int_0^1 r(x,v',z)dv'\right\} \\
&\cdot \left\{v^{-1}+2\int_0^1 r(x,v,z')dz'/z'\right\},
\end{aligned}
$$
$$x \ge 0. \tag{2.14}$$

In the limiting case of a slab of zero thickness overlying a complete absorber, the diffusely reflected intensity is

$$r(0,v,z) = 0. \tag{2.15}$$

This initial condition will be modified when there is a reflecting surface.

It is appropriate to introduce the function $S(x,v,z)$, which is symmetric in v and z, i.e.,

$$S(x,v,z) = S(x,z,v), \tag{2.16}$$

where

$$r(x,v,z) = S(x,v,z)/4v. \tag{2.17}$$

The Cauchy system for the function $S(x,v,z)$ is

$$
\begin{aligned}
S_x(x,v,z) &= -(z^{-1}+v^{-1})S(x,v,z) \\
&+ \lambda(x)\left\{1+\frac{1}{2}\int_0^1 S(x,v',z)dv'/v'\right\} \\
&\cdot \left\{1+\frac{1}{2}\int_0^1 S(x,v,z')dz'/z'\right\},
\end{aligned}
$$
$$x \ge 0, \tag{2.18}$$

$$S(0,v,z) = 0. \tag{2.19}$$

Let us next consider the intensity of diffusely transmitted light, and define $t(x,v,z)$ to be the intensity of the radiation diffusely transmitted with direction cosine v, by the slab of thickness x, with parallel ray illumination having direction cosine z and net flux π. See Fig. 2.4.

The function $T(x,v,z)$ is defined by the relation

$$t(x,v,z) = \frac{T(x,v,z)}{4v}. \tag{2.20}$$

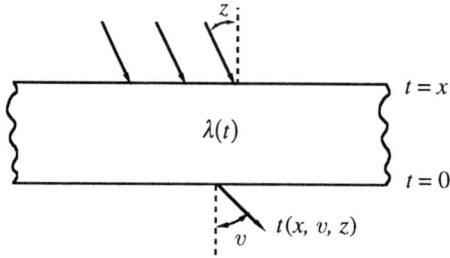

Figure 2.4. Diffusely transmitted radiation.

From either the physical (see Fig. 2.5) or analytical procedure, it can be shown that $T(x, v, z)$ satisfies the differential equation

$$
\begin{aligned}
T_x(x, v, z) \;=\;& -z^{-1} T(x, v, z) \\
& + \lambda(x) \left\{ 1 + \frac{1}{2} \int_0^1 S(x, v', z) dv'/v' \right\} \\
& \cdot \left\{ e^{-x/v} + \frac{1}{2} \int_0^1 T(x, v, z') dz'/z' \right\}, \\
& \hspace{5cm} 0 \le x \le x_1.
\end{aligned}
\tag{2.21}
$$

and the initial condition

$$
T(0, v, z) = 0.
\tag{2.22}
$$

Observe that the functional equation for T is coupled to that for S. The reflected intensity is the exact solution of the Cauchy system of Eqs. (2.18) and (2.19). To determine the transmission function, the enlarged Cauchy system of Eqs. (2.18), (2.19), (2.21) and (2.22) is to be solved.

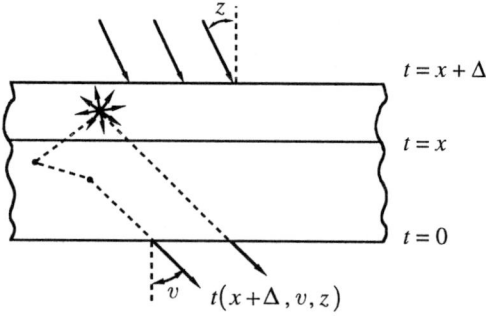

Figure 2.5. Transmission of radiation scattered into the direction $\arccos v$ either directly or after multiple scatterings, after scattering at the top.

2.3 Computational method and results

2.3.1 Discretization by Gaussian quadrature

In computing a solution of the Cauchy system (Eqs. (2.18), (2.19)) for reflection, we use a high order Gaussian quadrature formula of order N, and we obtain a system of ordinary differential equations suitable for solution by numerical integration.

In order to deal with the above initial value problem for integro-differential equations, we make the approximation that integrals can be replaced by sums according to the Gaussian quadrature formula [1–4],

$$\int_0^1 f(z)dz \cong \sum_{i=1}^{N} f(z_i)w_i, \tag{2.23}$$

where $z_1, z_2, \cdots, z_N$ are the points at which the integrand is evaluated, and $w_1, w_2, \cdots, w_N$ are the corresponding Christoffel weights. When we use a Gaussian quadrature formula of order N, the formula is exact for polynomials up to degree $2N - 1$. For example, for $N = 7$, the quadrature is exact if the polynomial is of degree 13; this is sufficiently accurate for many practical purposes. The points $z_1, z_2, \cdots, z_N$ are the roots of the shifted Legendre polynomial, $P_N(1 - 2x)$, which are in the interval $(0, 1)$. These roots and weights have been tabulated for various N. Table 2.1 gives them for $N = 7$ to eight significant figures. For other values of N, see Ref. [3]. Table 2.1 also gives the angles, arccos z_i.

The function S of three variables is discretized in the angles, and we introduce the functions

$$
\begin{aligned}
S_{ij}(x) &= S(x, z_i, z_j), \\
&\quad i = 1, 2, \ldots, N; \ j = 1, 2, \ldots, N.
\end{aligned}
\tag{2.24}
$$

The Cauchy system of Eqs. (2.18) and (2.19) is approximated by the system of ordinary differential equations,

$$
\begin{aligned}
S'_{ij}(x) &= -\left(z_i^{-1} + z_j^{-1}\right) S_{ij} \\
&\quad + \lambda(x)\left\{1 + 0.5\sum_{k=1}^{N} S_{kj}w_k/z_k\right\} \\
&\quad \cdot \left\{1 + 0.5\sum_{k=1}^{N} S_{ik}w_k/z_k\right\}, \\
&\qquad\qquad x \geq 0,
\end{aligned}
\tag{2.25}
$$

and the initial conditions,

$$S_{ij}(0) = 0.$$
$$i, j = 1, 2, \ldots, N. \tag{2.26}$$

Table 2.1. Points, Weights and Angles for Gaussian Quadrature with $N = 7$

$$\int_0^1 f(x)dz \cong \sum_{i=1}^N f(z_i)w_i$$

	Points	Weights	Angles(deg)
i	z_i	w_i	arccos z_i
1	.025446046	.064742484	88.5419
2	.12923441	.13985269	82.5746
3	.29707742	.19091502	72.7178
4	.50000000	.20897958	60.0000
5	.70292258	.19091502	45.3380
6	.87076559	.13985269	29.4523
7	.97455396	.064742484	12.9531

2.3.2 Numerical integration of a system of differential equations

Systems of ordinary differential equations with initial conditions may be solved numerically by extending the numerical methods of Chapter 1 [5], [6]. Instead of a scalar equation, consider the system of differential equations,

$$dr/dx = f(x, r), \quad r(c) = a = \text{given}, \quad c \leq x \leq C, \tag{2.27}$$

where $r(x)$ is a vector function of x whose components are $r_1(x), r_2(x), \ldots,$ $r_n(x)$; f is the vector function of x and the vector $r(x)$ whose components are the functions $f_1(x, r), f_2(x, r), \ldots, f_n(x, r)$; and a is the vector whose components are $a_1, a_2, \ldots, a_n$. In other words, the vector system (2.27) is equivalent to the n differential equations and initial conditions,

$$\begin{aligned} dr_1/dx &= f_1(x, r), \quad r_1(c) = a_1, \\ dr_2/dx &= f_2(x, r), \quad r_2(c) = a_2, \\ &\cdots \\ dr_n/dx &= f_n(x, r), \quad r_n(c) = a_n. \end{aligned} \tag{2.28}$$

Euler's method for a system of differential equations provides the formulas for the solution at the next $(i + 1)$st step of step size h:

$$\begin{aligned} r_1(i+1) &= r_1(i) + hf_1(x, r), \\ r_2(i+1) &= r_2(i) + hf_2(x, r), \\ &\cdots \\ r_n(i+1) &= r_n(i) + hf_n(x, r). \end{aligned} \tag{2.29}$$

where the vector r is evaluated at i, i.e., the vector $r(i)$. Observe that the updating of all the components of r is done "at the same time," i.e., in the same step of the cycle. This is stated by the above formulas using in the right hand sides the vector r evaluated at the ith step to compute the vector r at the $(i + 1)$st step.

The second-order Runge-Kutta, the fourth-order Runge-Kutta, and other numerical integration algorithms are likewise extended to systems of ordinary differential equations with no loss of generality.

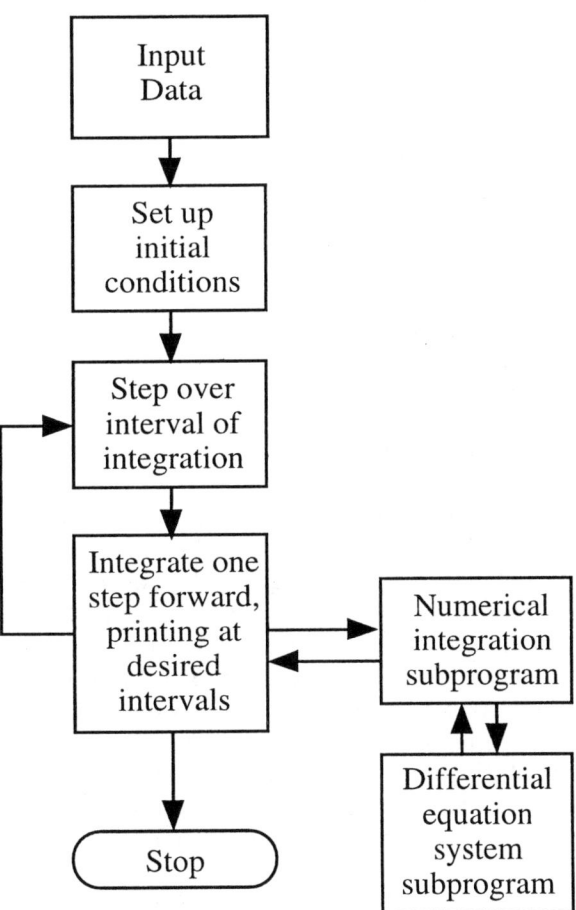

Figure 2.6. Flowchart for numerical integration.

The method of numerical integration of (2.28) is diagrammed in Figure 2.6. The first block, "input data," refers to the defining of the constants of the problem such as:

n = the number of differential equations, i.e., the dimension of the vector r;

c = the initial value of the independent variable, x;

h = the step size of integration;

m = the number of integration steps between $x = c$ and $x = C$;

a = the initial condition vector

and other quantities such as the number of steps between printing of the solution, etc.

The second block, "set up initial conditions," means that the initial conditions in Eqs. (2.28) are imposed.

In the flowchart of Fig. 2.6, "step over interval of integration" tells us to perform iteratively (by "looping") the algorithm of taking an integration step forward, whether the algorithm be Euler's, Runge-Kutta's, or some other one. Next, the block labelled, "integrate one step forward..." means that the appropriate algorithm is to be applied to all of the components during that step. If the step size is "large," then it may be appropriate to print after each integration step; otherwise, a certain number of steps (previously input or determined by another algorithm) are taken before printing.

During the stepping forward, the numerical integration algorithm is executed as represented by the block, "numerical integration subprogram," in a subprogram or its equivalent. The numerical integration subprogram is considered a "library" routine, since it is used in so many applications. The numerical integration algorithm requires that the right hand sides of the differential equations be computed; this is done in the box labelled, "differential equations system subprogram."

By following the flowchart, one can prepare a computer program for integrating a system of n differential equations with n initial conditions from $x = c$ to $x = C$ using a step size h. The program would contain a subprogram encoding the numerical integration algorithm. The user would provide the input data and must write the code for the system of differential equations, i.e., the evaluation of the right hand sides of the differential equations. Alternatively, one could use one of many commercially available software packages that replace the writing of the main program and integration subprograms; the differential equations are to be described as the software requires.

2.3.3 Computational procedure for reflection matrix

The computational procedure for determining the reflected intensities is to apply initial conditions in Eq. (2.26) and integrate Eqs. (2.25) from $x = 0$ to $x = $ the maximum value of interest, x_1. In the cases cited below, the maximum was 3.0, 10.0, 20.0 or 30.0, depending on the value of the albedo; the larger the albedo the larger the maximum thickness before saturation occurs. Note that the function $\lambda(x)$ is evaluated at x in the subprogram. A similar procedure holds for computing transmitted intensities.

In order to prepare the matrix differential equations for the reflection function whose components are R_{ij}, $i = 1, 2, \ldots, N$; $j = 1, 2, \ldots, N$, we must "vectorize" the matrix. We define the r vector as follows:

$$r_1 = R_{11},$$
$$r_2 = R_{12},$$
$$\ldots$$
$$r_N = R_{1N},$$
$$r_{N+1} = R_{21},$$
$$r_{N+2} = R_{22},$$
$$\ldots$$
$$r_{N^2} = R_{NN}.$$

We do this more succinctly in the Fortran "do loop":

$$k = 0$$
$$\text{do } 1 \ i = 1, N$$
$$\text{do } 1 \ j = 1, N$$
$$k = k + 1$$
$$1 \quad r(k) = R(i, j).$$

In special cases when the reflection matrix is symmetric, i.e., when

$$R(i, j) = R(j, i),$$

then only $N(N + 1)/2$ components are computed from the differential equations and the vector r has dimension $N(N + 1)/2$ and is defined from the R matrix through the loop,

$$k = 0$$
$$\text{do } 2 \ i = 1, N$$
$$\text{do } 2 \ j = 1, i$$
$$k = k + 1$$
$$2 \quad r(k) = R(i, j).$$

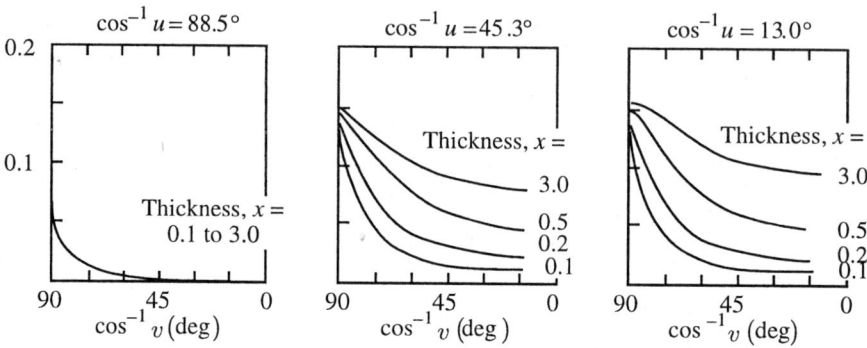

Figure 2.7. Reflected intensities for various thicknesses and three incident angles, albedo 0.5.

Observe that, in the inner loop, the index runs up to i, the index of the outer loop. The "uncomputed" components of the R matrix are determined through the identity given above.

2.3.4 Computational results

Extensive computations have been made for reflection and transmission functions of homogeneous and inhomogeneous slabs. The results were obtained after a Fortran program was written, checked, and executed; it uses a fourth-order Runge-Kutta method. The computed results were printed out to six significant digits and checked for numerical accuracy by varying the step size of integration, varying the order of the quadrature formula, and making comparisons against previously published results (mostly for special cases such as thin or semi-infinite slabs). It is believed that the results are accurate to five significant digits. We like to use odd values of N (3, 5, 7, 9, etc.) because the middle angle is always 60 degrees, and values can then be compared at 60 degrees. We used $N = 7$ in our production runs, which gives a good angular coverage. We generally use a step size of 0.01 and compare results with a step size of 0.005, using the 0.01 step for production runs. By "production runs" we mean the runs for producing extensive tables and graphs, carried out only after thorough checking has been completed.

We present selected tables of computed values so that you may check your work. The proper procedure is to prepare a computer program that reproduces these results. For this reason, we present tables containing about five significant figures. Until and unless your values agree with these, do not proceed with further cases. Do not be satisfied if they agree only to two or

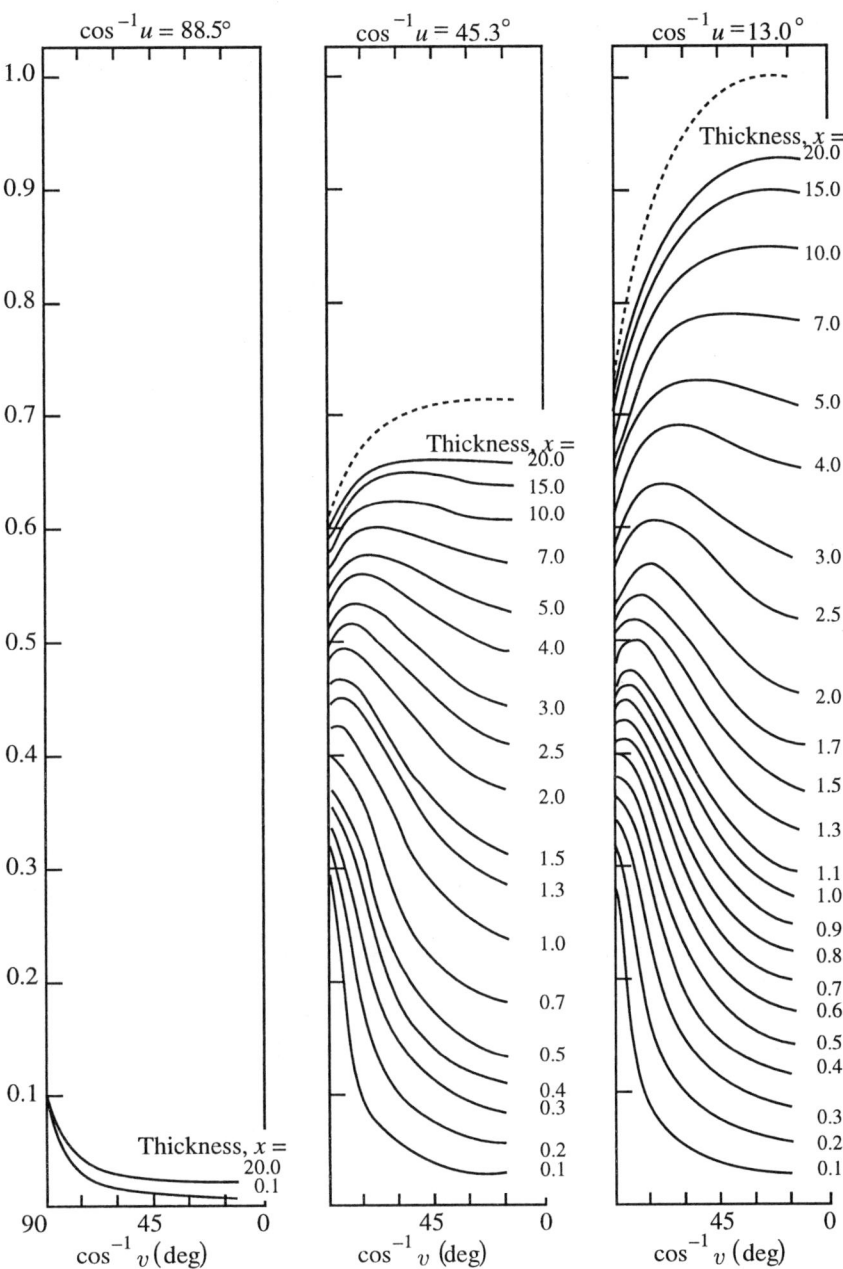

Figure 2.8. Reflected intensities for various thicknesses and three incident angles, albedo 1.0.

three digits; modern digital computers are much more accurate than that. If you obtain only mild agreement, this indicates that there is an error in the program—not a gross error which is easy to detect (such as giving rise to negative intensities), but the more difficult subtle error or errors.

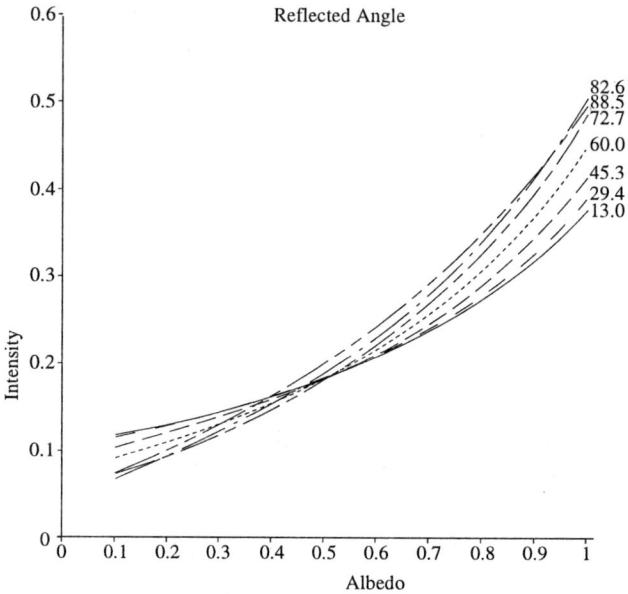

Figure 2.9. Reflected intensities as functions of albedo, for incident angle 13.0°, thickness 1.0.

Tables 2.2–2.4 have values of reflected intensities to six decimal places and can be used to confirm one's computational results. The intensities in the rows are for the seven incident angles 88.5 through 13.0 degrees given in Table 2.1. The columns are for reflected angles 88.5 through 13.0 degrees, respectively, from left to right. For example, with incident angle 13.0 degrees, the reflected intensity emerging at 60 degrees from the normal is 0.017619.

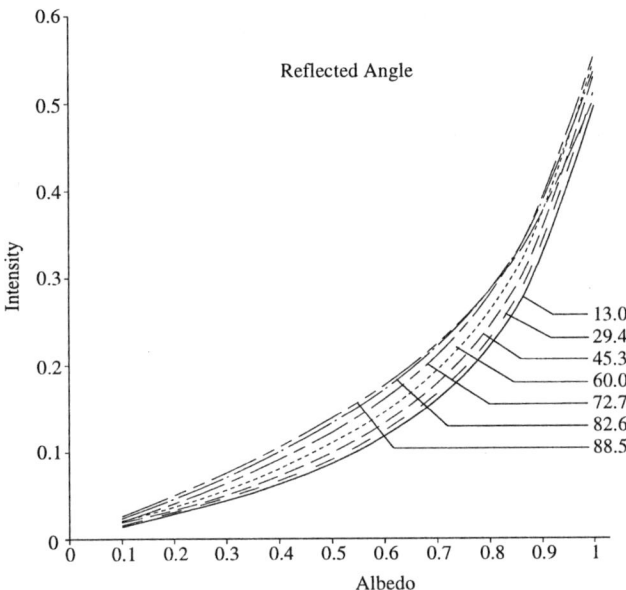

Figure 2.10. Reflected intensities as functions of albedo, for incident angle 13.0°, thickness 3.0.

Table 2.2. Reflected Intensities for Albedo 0.1, Thickness 3.

Incident Angle	Reflected Angle (degrees)						
	88.5	82.6	72.7	60.0	45.3	29.4	13.0
88.5	.012619	.004192	.002027	.001252	.000906	.000738	.000663
82.6	.021291	.012865	.007865	.005360	.004068	.003393	.003078
72.7	.023667	.018079	.013079	.009807	.007847	.006735	.006192
60.0	.024593	.020736	.016505	.013233	.011043	.009713	.009039
45.3	.025038	.022128	.018566	.015525	.013332	.011935	.011207
29.4	.025267	.022864	.019740	.016915	.014784	.013382	.012638
13.0	.025373	.023211	.020314	.017619	.015537	.014144	.013397

Reflected intensities. Graphs are presented for reflected and transmitted intensities for selected albedoes, for three angles of incidence, and for homogeneous slabs of various thicknesses, and showing their dependence on the reflected angle. The dependence of reflected intensity on the angle of reflection is shown for slabs of different thicknesses with albedo 0.5 in Fig. 2.7 and albedo 1.0 in Fig. 2.8. In Fig. 2.7, the curves only go out to thickness 3.0 since there is very little change at this thickness. In other words, a thickness of 3.0 may be considered to be semi-infinite, a case which has been studied by classical means. When the slab scatters radiation conservatively, all absorbed radiation is reemitted, so that the intensity increases with increasing optical thickness. In Fig. 2.7, the curves for incident angle 13 degrees keep growing

Table 2.3. Reflected Intensities for Albedo 0.5, Thickness 3.

Incident Angle	Reflected Angle (degrees)						
	88.5	82.6	72.7	60.0	45.3	29.4	13.0
88.5	.065832	.022918	.011570	.007379	.005462	.004511	.004075
82.6	.116396	.073704	.047037	.033111	.025691	.021722	.019840
72.7	.135075	.108127	.081663	.063245	.051727	.044999	.041661
60.0	.144985	.128103	.106445	.088146	.075176	.067010	.062788
45.3	.150890	.139739	.122392	.105686	.092741	.084114	.079509
29.4	.154351	.146358	.131896	.116699	.104199	.095536	.090808
13.0	.156084	.149610	.136666	.122381	.110234	.101631	.096877

Table 2.4. Reflected Intensities for Albedo 1.0, Thickness 3.

Incident Angle	Reflected Angle (degrees)						
	88.5	82.6	72.7	60.0	45.3	29.4	13.0
88.5	.143042	.055872	.032203	.023096	.018665	.016312	.015181
82.6	.283759	.201481	.146654	.115932	.098052	.087629	.082378
72.7	.375962	.337120	.289876	.251507	.223633	.205193	.195277
60.0	.453814	.448533	.423302	.391432	.361672	.339046	.326000
45.3	.515589	.533317	.529144	.508455	.481587	.457990	.443466
29.4	.558186	.590433	.601442	.590459	.567348	.544137	.529081
13.0	.581406	.621214	.640600	.635409	.614835	.592144	.576947

asymptotically to the dashed curve representing a semi-infinite medium. The calculations were stopped at thickness 20.0 after 2,000 integration steps.

In each of the next three figures, the family of curves represents seven angles of reflection. The dependence of reflected intensity on albedo is shown in Fig. 2.9 for thickness 1.0, while the same is shown for thickness 3.0 in Fig. 2.10. Reflected intensity increases with albedo, as one would expect. However, these curves cross in an interesting manner when the thickness is 3.0. Fig. 2.11 shows reflected intensity as a function of thickness when the albedo is 1.0. Here, too, the curves cross each other.

Transmitted intensities. Examination of the transmitted intensity curves in Figs. 2.12 and 2.13 reveals similar interesting patterns. Most obvious, perhaps, is the exponential dropoff of intensities with increasing thickness, for all albedoes less than 1.0. The conservative case of 1.0 is, again, very different from the rest.

Layered medium. Consider an inhomogeneous slab of unit optical thickness whose albedo for single scattering 0.4 is in the lower layer and 0.6 in the upper layer, with a rapidly changing transition layer at the midpoint. A continuous function that displays this property is the function,

$$\lambda(t) = 0.5 + 0.1 \tanh 10(t - 0.5), \qquad 0 \le t \le 1.0.$$

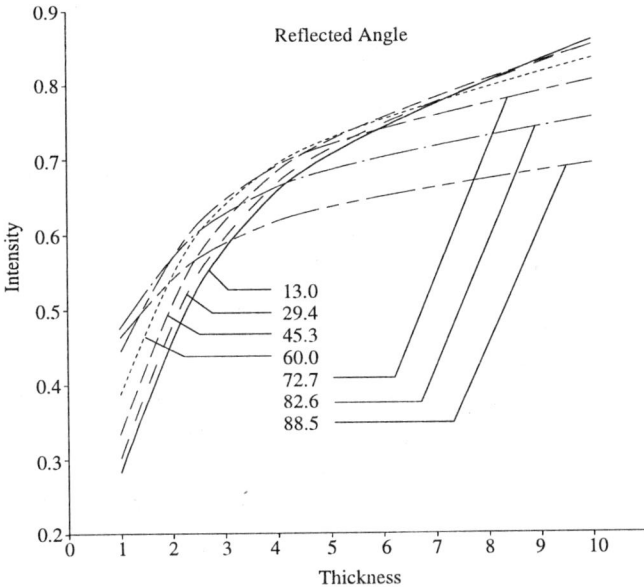

Figure 2.11. Reflected intensities as functions of thickness, for incident angle 13.0°, albedo 1.0.

In computing the solution of the Cauchy problem, one must be certain to take integration steps that are sufficiently small over the transition region. The reflected intensity patterns for three incident directions are shown in Fig. 2.14. This graph can be compared with that of Figs. 2.7 and 2.8. This is only illustrative of some of the capabilities of the computer program for reflection from a vertically inhomogeneous medium.

In the following chapter, the question of estimating parameters such as albedo functions, optical thickness, angle of incidence, and others, based on noisy measurements of reflected or transmitted intensities, will be addressed.

2.4 Internal Intensity and Source Functions

We continue our study of radiation fields in inhomogeneous plane-parallel media illuminated by uniform parallel rays of radiation. We turn our attention now to the determination of internal intensity and source functions.

2.4.1 Basic Cauchy problem

We define the internal intensity function, sketched in Fig. 2.15,

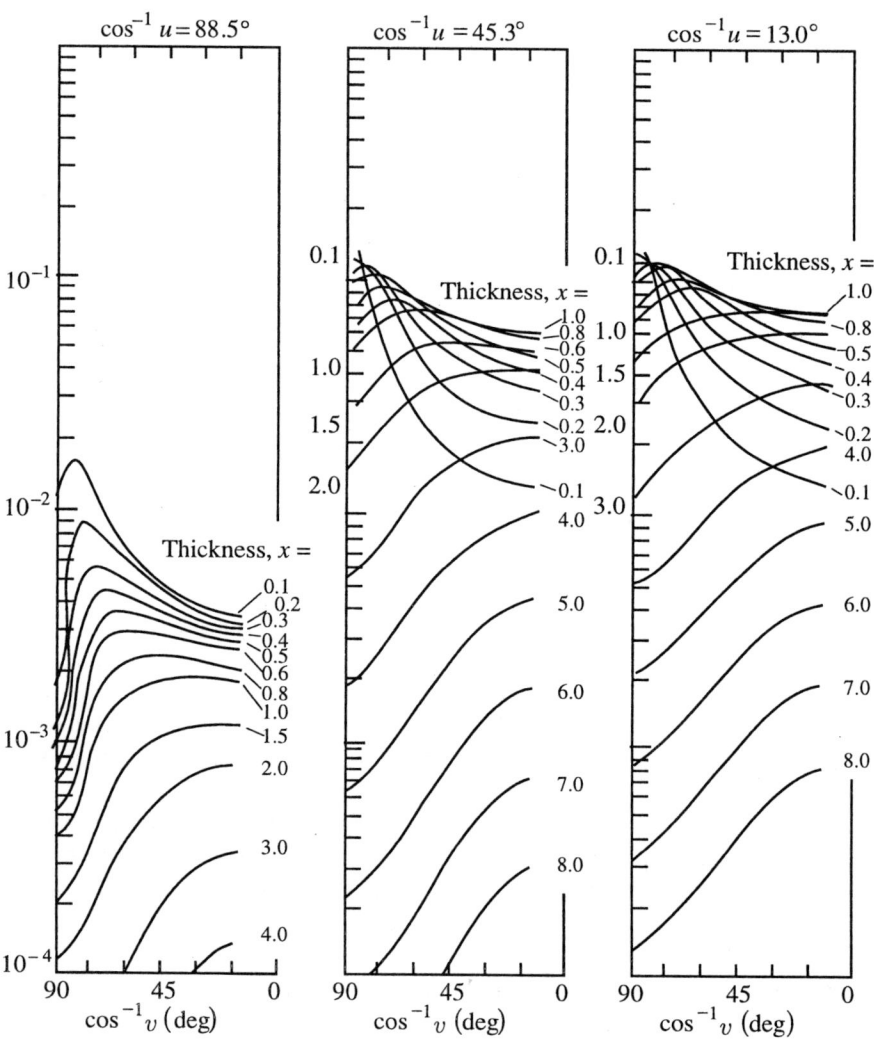

Figure 2.12. Transmitted intensities for various thicknesses and three incident angles, albedo 0.5.

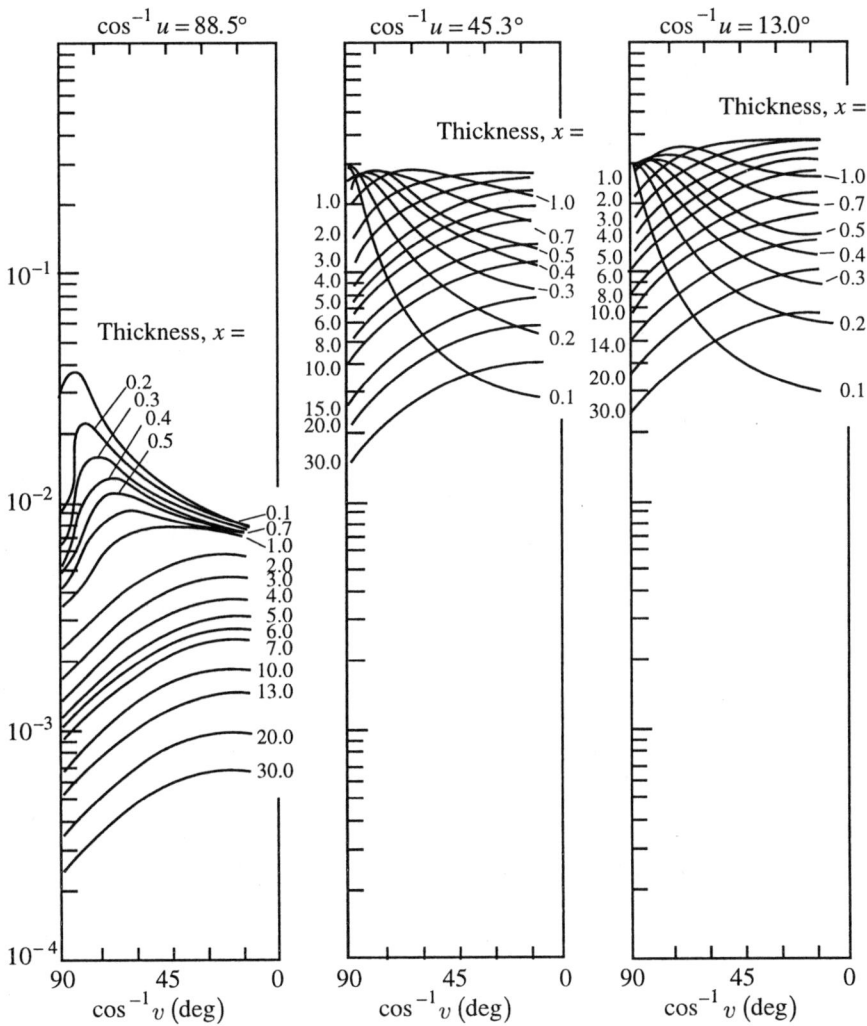

Figure 2.13. Transmitted intensities for various thicknesses and three incident angles, albedo 1.0.

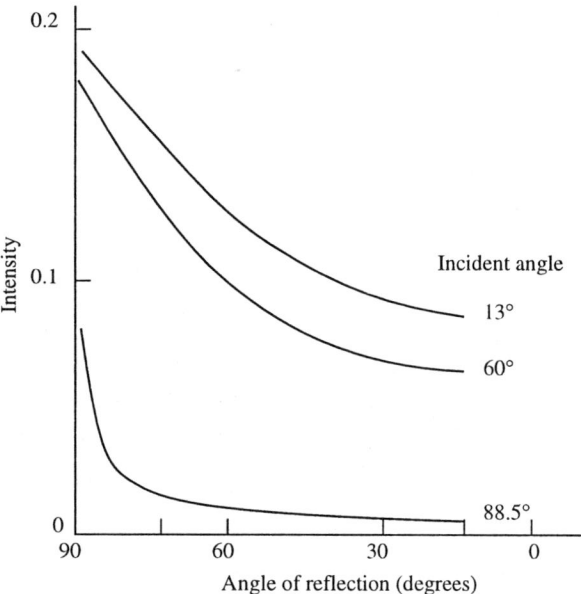

Figure 2.14. Reflected intensities for a layered medium. ($\lambda(t) = 0.5 + 0.1\tanh 10(t - 0.5)$, thickness= 1.0).

$I(t, v, x, u)$ = the intensity of the diffuse radiation at alti- (2.30)
tude t propagating with direction cosine v, in
an inhomogeneous medium of thickness x due
to uniform parallel radiation at the top with
incident direction cosine u.

Observe that directions are specified by v, the cosine of the angle measured from the upward vertical. Upwelling directions have $0 \leq v \leq 1$, while down-welling directions have $-1 \leq v \leq 0$, also expressed as $-v$ where $0 \leq v \leq 1$.

Invariant imbedding equations are obtained in a manner analogous to that employed in Section 1.3 for the rod problem. We add a thin layer of thickness Δ to the top of the slab of thickness x. We enumerate all the changes in internal intensity due to the presence of this layer. We have

$$I(t, v, x + \Delta, u) - I(t, v, x, u) =$$
$$- \Delta u^{-1} I(t, v, x, u)$$
$$+ 2\Delta J(x, x, u) \int_0^1 I(t, v, x, u') du'/u'$$
$$+ \Delta g(t, v, x, u), \qquad (2.31)$$

where the function g is introduced as follows

$$g(t, v, x, u) = \begin{cases} 0, & 0 \leq v \leq 1, \\ J(x, x, u)(-v^{-1}) \exp(+(x - t)/v), & -1 \leq v \leq 0, \end{cases} \quad (2.32)$$

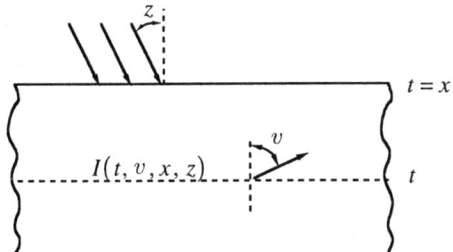

Figure 2.15. The internal intensity of the diffuse radiation.

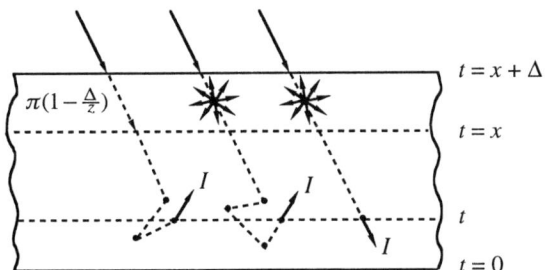

Figure 2.16. A thin layer is added to the top of the slab of thickness x.

The first equation is obtained by adding a thin slab of thickness Δ to the top of a slab of thickness x, accounting for all the changes in the internal intensity due to the presence of this added slab, then letting the thickness of the thin slab go to zero and $J(x, x, u)$ is the rate of production of scattered radiation at the top, given by expression (2.10). The first term on the right hand side of Eq. (2.31) is due to the diminution in I due to diminution of incoming radiation while the second is due to an added rate of production of scattered radiation at the top. Observe that this scattering is equivalent to having uniform isotropic sources at the top which produce diffuse intensities at altitude t; hence there is an integral over all incoming directions. For upwelling intensities, there is no other term in the differential equation. On the other hand, for downwelling intensities, we must account for the rate of production of radiation at the top which directly, i.e., only by attenuation along the path, gives rise to diffuse radiation at the point in question, as expressed by (2.32). These processes are represented in Fig. 2.16.

We let the thickness of the added thin layer go to zero. We obtain the exact invariant imbedding equations as an integro-differential equation with initial condition:

$$
\begin{aligned}
I_x(t, v, x, u) \; = \; & -u^{-1} I(t, v, x, u) + \\
& + 2J(x, x, u) \int_0^1 I(t, v, x, u') du'/u' \\
& + g(t, v, x, u), \\
& t \le x, \; -1 \le v \le 1, \; 0 \le t, \; 0 \le u \le 1, \qquad (2.33)
\end{aligned}
$$

where

$$
g(t, v, x, u) = \begin{cases} 0, & 0 \le v \le 1, \\ J(x, x, u)(-v^{-1}) \exp(+(x-t)/v), & -1 \le v \le 0, \end{cases} \qquad (2.34)
$$

and the initial conditions which are appropriate when the thickness x has the value t are:

$$
I(t, v, t, u) = \begin{cases} r(t, v, u), & 0 \le v \le 1, \\ 0 & -1 \le v \le 0. \end{cases} \qquad (2.35)
$$

The reflection function is given by the Cauchy problem of Eqs. (2.18) and (2.19) in Section 2.2.2. For the rate of production of scattered radiation at the top, we use the expression (from (2.10)),

$$
J(x, x, u) = \frac{1}{2} \left\{ 1 + (1/2) \int_0^1 S(x, v, u) dv/v \right\}. \qquad (2.36)
$$

Next, the invariant imbedding equations will be obtained for the source function, $J(t, x, u)$ at altitude t in the inhomogeneous medium of thickness x.

We define the source function,

$$
\begin{aligned}
J(t, x, u) \; = \; & \text{The rate of production of scattered radiation,} \qquad (2.37) \\
& \text{per unit volume per unit solid angle per unit} \\
& \text{time per unit frequency interval, at altitude} \\
& t \text{ in a slab of thickness } x \text{ due to illumination} \\
& \text{with direction cosine } u.
\end{aligned}
$$

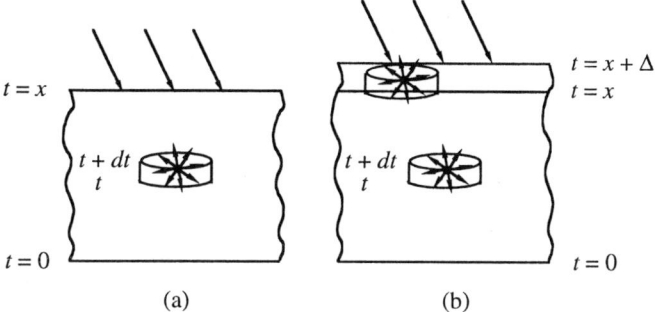

Figure 2.17. Invariant imbedding for the source function. (a) A slab of thickness x. (b) A slab of thickness $x + \Delta$.

The source function at altitude t in a slab of thickness x is sketched in Fig. 2.17(a). The rate of production in a volume with unit horizontal cross-sectional area and height Δ, located between altitudes t and $t + dt$ is $J(t, x, u)dt$. In Fig. 2.17(b) is sketched the source function in a slab of thickness $x + \Delta$, and the rate of production in an identical volume is $J(t, x + \Delta, u)dt$. We can express the latter in the following way:

$$J(t, x + \Delta, u)dt \;=\; J(t, x, u)dt - (\Delta/u)J(t, x, u)dt \qquad (2.38)$$

$$+J(x, x, u)\Delta(1/\pi)(2\pi)\int_0^1 J(t, x, u')dtdu'/u'$$

$$+o(\Delta).$$

The negative term is due to the reduction of incoming radiation which illuminates the slab of thickness x and produces the source function as shown above, and the additive term is the contribution of the rate of production (in units of π) at the top into all directions u' which induces production at altitude t.

Now, dividing through by Δ and letting the increment $\Delta \to 0$, we obtain the differential-integral equation,

$$J_x(t, x, u) \;=\; -u^{-1}J(t, x, u) + 2J(x, x, u)\int_0^1 J(t, x, u')du'/u'$$

$$0 \leq x \leq x_1. \qquad (2.39)$$

The rate of production at the top, in turn, may be expressed as the sum of the radiation that was emerging upward from the slab of thickness x that interacts in this added small layer and causes more scattering to take place there, namely

$$J(x, x, u) = \frac{\lambda(x)}{4} \left\{ 1 + 2 \int_0^1 r(x, v, u) dv \right\}, \qquad (2.40)$$

or,

$$J(x, x, u) = \frac{\lambda(x)}{4} \left\{ 1 + \frac{1}{2} \int_0^1 S(x, v, u) dv/v \right\}, \qquad (2.41)$$

where $r(v, x, u)$ is the intensity of diffusely reflected radiation. The two terms on the right are the incident radiation interacting and scattering at the top, and the radiation that would be emerging with direction cosine v interacting and scattering at the top,

$$J(t, t, u) = \frac{\lambda(t)}{4} \left\{ 1 + \frac{1}{2} \int_0^1 S(t, v, u) dv/v \right\}. \qquad (2.42)$$

Observe that the necessary equations include those for reflected intensities for thickness $x \geq 0$, namely (2.7) and (2.8). When internal intensities are desired at altitude $t(0 \leq t \leq x)$, then initial condition (2.35) is imposed and Eq. (2.33) is adjoined to the differential equations for reflected intensities. The enlarged system of differential equations must then be solved until x has the desired thickness, x_1. The same can be said of the evaluation of the source function using (2.39) and (2.41).

2.4.2 Computational method

The above Cauchy problem is approximated to a high degree of precision to obtain a system of ordinary differential equations with a complete set of initial conditions. The Gaussian quadrature method with N points, $z_1, z_2, \ldots, z_N$, is used for the internal intensities. Fix the value of t, the altitude of interest. Introduce the N upwelling internal intensities as functions of x,

$$U_{ij}(x) = I(t, z_i, x, z_j), \qquad (2.43)$$

where i and j take on the integer values $1, 2, \ldots, N$, and let

$$\begin{aligned} A_j(x) &= J(x, x, z_j) \\ &= \frac{1}{2} \left\{ 1 + \frac{1}{2} \sum_{k=1}^N S_{jk}(x) w_k / z_k \right\} \end{aligned} \qquad (2.44)$$

Then the differential-integral equation becomes a system of ordinary differential equations

$$U'_{ij} = -z_j^{-1} U_{ij} + \left\{ 1 + \frac{1}{2} \sum_{k=1}^{N} S_{kj}(x) w_k / z_k \right\}$$

$$\cdot \sum_{k=1}^{N} U_{ik} w_k / z_k, \tag{2.45}$$

with initial conditions,

$$U_{ij}(t) = r_{ij}(t) = S_{ij}(t)/4z_i. \tag{2.46}$$

The downwelling intensities are

$$V_{ij}(x) = I(t, -z_i, x, z_j) \tag{2.47}$$

and they satisfy the differential equation system

$$V'_{ij} = -z_j^{-1} V_{ij}$$

$$+ \frac{1}{2} \left\{ 1 + (1/2) \sum_{k=1}^{N} S_{kj}(x) w_k / z_k \right\}$$

$$\cdot \left\{ 2 \sum_{k=1}^{N} V_{ik} w_k / z_k + z_i^{-1} \exp(-(x-t)/z_i \right\},$$

$$t \le x \le x_1 \tag{2.48}$$

with initial conditions

$$V_{ij}(t) = 0. \tag{2.49}$$

These equations hold for any value of $t \le x$ and for x ranging from zero to the desired final thickness, x_1.

The equations for $J(t, x, u)$ are obtained by introducing

$$J_j(x) = J(t, x, z_j), \qquad j = 1, 2, \ldots, N. \tag{2.50}$$

The equations are

$$J'_j = -z_j^{-1} J_j + \frac{1}{2} \left\{ 1 + (1/2) \sum_{k=1}^{N} S_{kj}(x) w_k / z_k \right\}$$

$$\cdot \left\{ 2 \sum_{k=1}^{N} J_k w_k / z_k \right\}, \tag{2.51}$$

$$t \le x \le x_1,$$

$$J_j(t) = \frac{1}{2}\left\{1 + (1/2)\sum_{k=1}^{N} S_{kj}(t)w_k/z_k\right\}$$

$$j = 1, 2, \ldots, N. \tag{2.52}$$

The computational procedure is to integrate the differential equations for $S_{ij}(x)$ from $x = 0$ to $x = t$. Then adjoin the $2N^2 + N$ differential equations for U_{ij}, V_{ij}, and J_j and use the initial conditions as given above. The enlarged system of differential equations is integrated to $x =$ the desired final thickness, x_1. Observe that solutions are obtained for internal intensities at altitude t for any thickness x greater than or equal to t. The procedure, depicted in Fig. 2.18, readily produces solutions for any set of values of t.

Let's take an example of the computation of the source function for albedo 0.8 at altitudes 0, 4, 8, ..., x, for values of x being 0, 4, 8, ..., 20. We choose $N = 7$ and a step size of 0.01. We have a system of $N(N + 1)/2 = 7(8)/2 = 28$ differential equations for the discretized S function to start the Cauchy calculations. We add 7 differential equations for the discretized source function J each time that x reaches the t values 4, 8, 12, 16, and 20. When $x = 0$, there are $28 + 7 = 35$ differential equations in our system; when $x = 4$, there are $35 + 7 = 42$ equations, and so forth. We output entries in a table of the J function whenever x is 4, 8, 12, 16, and 20. This means that there are $4.0/0.01 = 400$ integration steps between printouts. (Normally, we like to go 100 steps between outputting computed values.) The table has the form shown below in Table 2.5. We remark that the bottom row of each block for a given thickness contains the newly added seven values of the source function. The values for altitude $t = 0$ in all thicknesses have resulted from differential equations 29 through 35, and these seven differential equations were integrated 2,000 steps. The values for altitude $t = 4$ (equations 36 through 42) were initialized at $x = 4$ and the seven differential equations were integrated from 4 to 20, a total of 1600 steps. Although more equations are being adjoined, the number of remaining integration steps keeps decreasing. When the computations were completed at $x = 20$, 35 differential equations had been adjoined to the original 35. This is a very modest computational effort to obtain this quantity of information.

2.4.3 Computational results

Extensive numerical results have been obtained in the form of tables and graphs. Selected graphs are presented in Figs. 2.19–2.22. Figure 2.19 shows internal intensity fields in a slab of thickness 2.0 with an albedo of 0.5, while Fig. 2.20 does the same for a slab of thickness 10.0 with an albedo of 1.0. The incident angle is 60 degrees, or a direction cosine of 0.5. Study of the collection of results (see also *Multiple Scattering Processes* by Kagiwada, Kalaba, and Ueno, Ref. [7], Chapter 1), indicates the following: for albedoes less than one, the drop in intensity as a function of optical depth is exponential, while for an albedo of one the drop is linear.

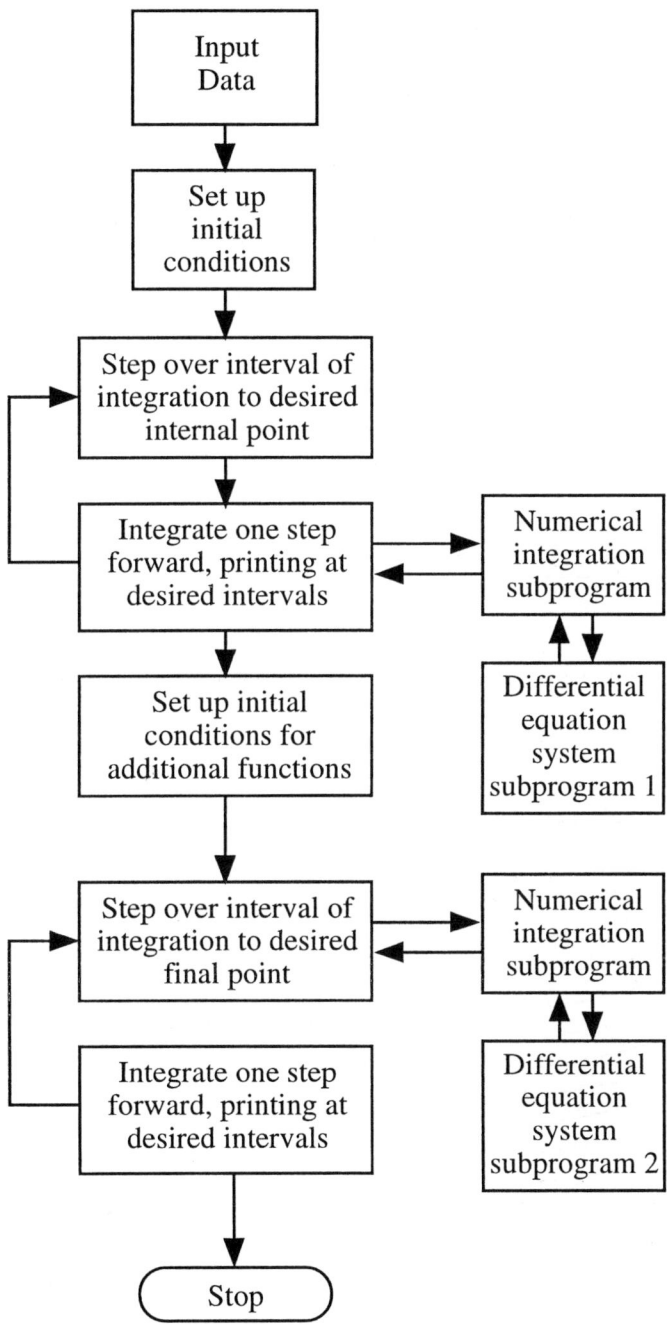

Figure 2.18. Flowchart for numerical integration with additional differential equations.

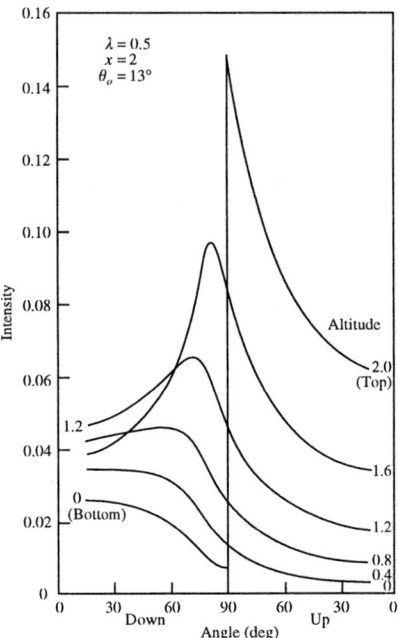

Figure 2.19. Internal intensities for albedo 0.5, thickness 2, incident direction cosine 0.5 (60 degrees).

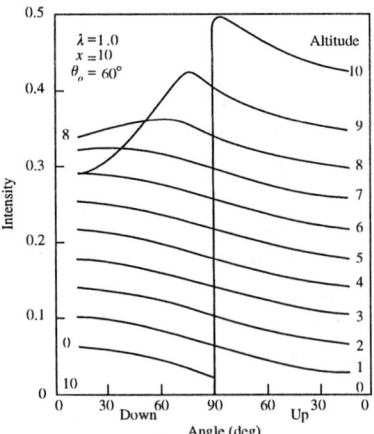

Figure 2.20. Internal intensities for albedo 1.0, thickness 10, incident direction cosine 0.5 (60 degrees).

Table 2.5. Source Function for Albedo 0.8

Thickness x	Angle (deg): Altitude t	88.5 $J1$	82.6 $J2$	72.7 $J3$	60.0 $J4$	45.3 $J5$	29.5 $J6$	13.0 $J7$
0.0	0.0	0.0	0.0	0.0	0.0	0.0	0.0	0.0
4.0	0.0	xxx	xxx	xxx	xxx	xxx	xxx	xxx
	4.0	xxx	xxx	xxx	xxx	xxx	xxx	xxx
8.0	0.0	xxx	xxx	xxx	xxx	xxx	xxx	xxx
	4.0	xxx	xxx	xxx	xxx	xxx	xxx	xxx
	8.0	xxx	xxx	xxx	xxx	xxx	xxx	xxx
...	...	...	...	...	...	...	...	...
20.0	0.0	xxx	xxx	xxx	xxx	xxx	xxx	xxx
	4.0	xxx	xxx	xxx	xxx	xxx	xxx	xxx
	8.0	xxx	xxx	xxx	xxx	xxx	xxx	xxx
	12.0	xxx	xxx	xxx	xxx	xxx	xxx	xxx
	16.0	xxx	xxx	xxx	xxx	xxx	xxx	xxx
	20.0	xxx	xxx	xxx	xxx	xxx	xxx	xxx

For source functions we show Fig. 2.21 for slabs of thicknesses 4, 8 and 20 and albedo 0.8, and Fig. 2.22 for slabs of thicknesses 2, 4, 8, 20 and 30 and albedo 1.0. In these, three angles of incidence are represented, 13.0, 60.0, and 88.5 degrees. In order to plot the nine curves of Fig. 2.21, we used a table of source functions like that described in section 2.4.2. For the three curves labelled 13.0 degrees angle of incidence, we looked at column 7 in the table. For a thickness of 4, we plotted the values of the source function at altitude 4 (at the depth 0) and altitude 0 (at depth 4). Next, for thickness 8, we plotted the three values of the source function at altitudes 0, 4 and 8 at the depths 8, 4, and 0, respectively. We used the same procedure for thickness 20, and we repeated the entire process for the two other incident angles in the figure.

In Figs. 2.21 and 2.22 we have plotted the source function against optical *depth* instead of altitude, since we wanted to show how the rate of production drops off as the observation point moves down into the medium, away from the incident radiation. (Depth is thickness minus altitude.) Observe that in Fig. 2.21, the three curves for incident angle 13.0 degrees overlap each other: the curve ending at depth 4 is the curve for thickness 4, depth 8 is for thickness 8, and depth 20 is for thickness 20. There is a small diminution at the lower boundary due to the absorbing barrier. The same comments hold for 60.0 and 88.5 degrees of incidence. In Fig. 2.22, five optical thicknesses are represented, and the albedo is 1.0. These curves remain at relatively high values due to the conservative scattering which keeps photons active, especially when the medium is thick.

2.5 Internal Emitting Sources

In many planetary and terrestrial media, there are vertically distributed sources of radiation emitting at the rate

$$g(t) \quad = \quad \text{emitted energy per unit volume per unit solid} \quad (2.53)$$
$$\text{angle per unit time at altitude } t,$$
$$0 \le t \le x.$$

The invariant imbedding method produces Cauchy problems for the emergent intensity of radiation, source function, and internal intensity functions appropriate for these distributed sources. The medium is assumed to scatter radiation isotropically with an albedo for single scattering that varies with altitude. In what follows, the expression "with direction cosine v" is used to indicate that the direction of propagation has a polar angle whose cosine is v.

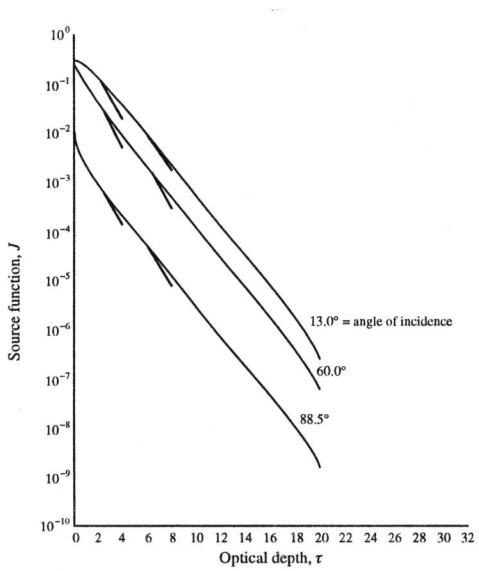

Figure 2.21. Nine source functions for three incident angles and three thicknesses (4, 8, and 20), albedo 0.8.

2.5.1 Emergent intensity

Let us introduce the emergent intensity function,

$$e(v, x) \quad = \quad \text{intensity of radiation emerging from the top} \quad (2.54)$$
$$\text{with direction cosine}$$
$$v, \quad 0 < v \le 1.$$

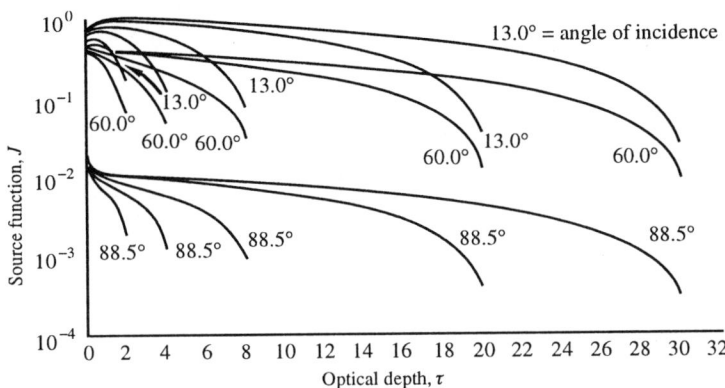

Figure 2.22. Fifteen source functions for three incident angles and five thicknesses (2, 4, 8, 20 and 30), albedo 1.0.

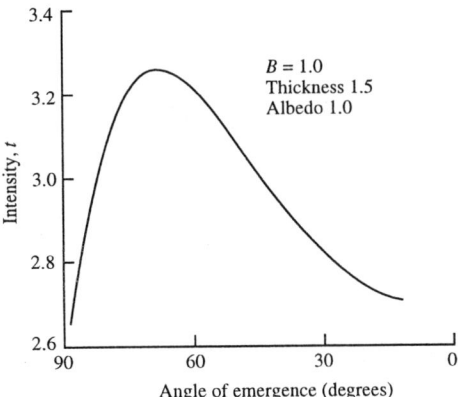

Figure 2.23. Emergent intensity for a slab as a function of angle of emergence with albedo 1.

We see from the previous discussions that a function of interest is

$$
\begin{aligned}
X(x, v) \quad = \quad & \text{fraction of energy scattered at the top that,} \\
& \text{emerges from the top with direction cosine } v, \\
& 0 \le v \le 1 \\
= \quad & \frac{1}{2}\left[1 + \frac{1}{2}\int_0^1 S(x, v, u)du/u \right].
\end{aligned} \qquad (2.55)
$$

Carrying out an invariant imbedding operation of adding a thin slab to the top, we obtain the equation,

$$
e(v, x + \varDelta) = e(v, x) - v^{-1}e(v, x)\varDelta + v^{-1}X(x, v)u(x, x)\varDelta + o(\varDelta), \qquad (2.56)
$$

where $u(x, x)$ is the source function at the top,

$$
u(x, x) = g(x) + \frac{1}{2}\int_0^1 e(v, x)dv. \qquad (2.57)
$$

After the limiting operation, we obtain the differential-integral equation,

$$
\begin{aligned}
e_x(v, x) \quad = \quad & -v^{-1}e(v, x) \\
& + v^{-1}\frac{1}{2}\left\{ 1 + \frac{1}{2}\int_0^1 S(x, v, u)du/u \right\} \\
& \cdot \left\{ g(x) + \frac{\lambda(x)}{2}\int_0^1 e(v, x)dv \right\}, \\
& 0 \le x \le x_1
\end{aligned} \qquad (2.58)
$$

The initial condition is

$$
e(v, 0) = 0, \qquad (2.59)
$$

assuming that the lower surface is a perfect absorber. To these equations must be adjoined the Cauchy problem for the reflection function, given earlier in this chapter.

2.5.2 Computational results for emergent intensities

Various cases were studied with different distributions of sources. In Fig. 2.23 we show the emergent intensity for a slab of thickness 1.5 with albedo 1.0 and a uniform distribution of sources, $g(t) = 1.0$. Next, consider seven different slabs, each having a layer of sources of unit strength. The first slab, Slab 1, has sources in the bottom one-seventh of the slab, and none elsewhere, so that the distribution function is

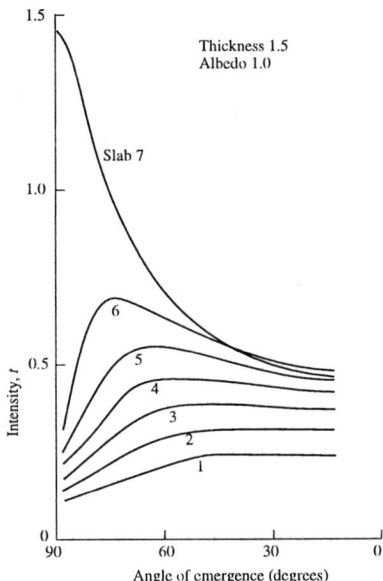

Figure 2.24. Emergent intensities for slabs 1–7 as a function of angle of emergence.

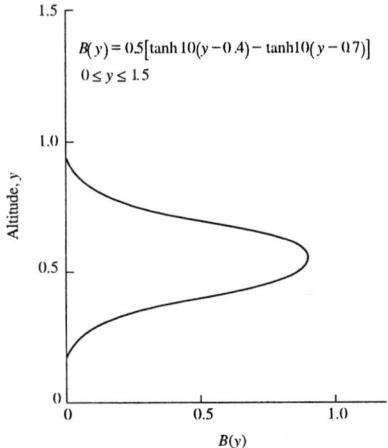

Figure 2.25. Source distribution function.

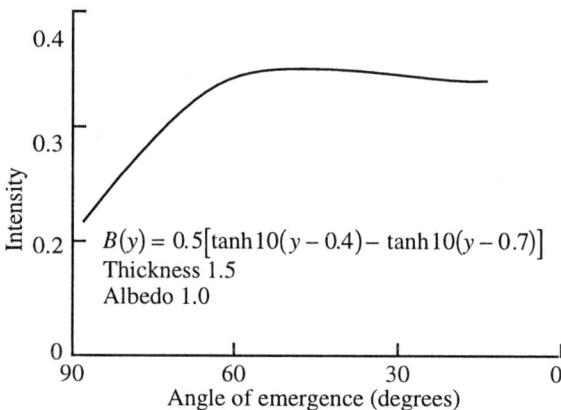

Figure 2.26. Emergent intensities for a slab as a function of angle of emergence.

$$g(t) \quad = 1.0, \text{ for } 0 \le y < (1/7)(1.5),$$
$$= 0.0 \text{ elsewhere.}$$

In Slab 2, the sources are in the next one-seventh portion, and so forth. With these seven distributions, the emergent intensities are as shown in Fig. 2.24. The seven curves represent the seven slabs. When the emitting sources are at the bottom, the emergent intensities are rather small. As we move the sources upward, the intensities increase. The sources at the top give rise to intensities that drop a bit near normal emergence and increase sharply toward the grazing direction.

Another distribution has a peak at altitude 0.5 in the slab of thickness 1.5. This is plotted in Fig. 2.25, and the emergent intensity is shown in Fig. 2.26. The intensity curve in Fig. 2.26 resembles that of Slab 3 in Fig. 2.24. Studies have been performed in which the distribution is perturbed and the effect examined. See [12] for additional details.

2.5.3 Source function

Let us introduce the source function,

$$U(t,x) \quad = \quad \text{rate of production of energy per unit volume} \quad (2.60)$$
$$\text{at altitude } t \text{ in a slab of thickness } x \text{ due to an}$$
$$\text{emitting source distribution } g(y), 0 \le t, y \le x.$$

Recall the source functions,

$$J(t, x, u) \quad = \quad \text{rate of production of energy per unit volume,} \quad (2.61)$$
at altitude t in a slab of thickness x due to uniform parallel illumination at the top with direction cosine $u, 0 \le t \le x; 0 \le u$

and

$$\Phi(t, x) \quad = \quad \text{rate of production of energy per unit volume}$$
at altitude t in a slab of thickness x due to uniform isotropic illumination at the top, $0 \le t \le x$

$$= \quad \int_0^1 \frac{1}{\pi} v J(t, x, v) 2\pi dv \qquad (2.62)$$

$$= \quad 2 \int_0^1 J(t, x, v) dv / v.$$

The finite invariant imbedding equation is

$$U(t, x + \Delta) = U(t, x) + U(x, x)\Delta\Phi(t, x) + o(\Delta), \qquad (2.63)$$

where $U(x, x)$ and $\Phi(t, x)$ are given above. This leads to the integro-differential equation,

$$U_x(t, x) \quad = \quad \left\{ g(x) + \frac{\lambda(x)}{2} \int_0^1 e(v, x) dv \right\}$$
$$\cdot \left\{ 2 \int_0^1 J(t, x, u) du / u \right\},$$
$$t \le x \le x_1 \qquad (2.64)$$

and the initial condition

$$U(t, t) = g(t) + \frac{\lambda(t)}{2} \int_0^1 e(v, x) dv, \ 0 \le t. \qquad (2.65)$$

The Cauchy problem for the source function $J(t, x, u)$ must be adjoined.

2.5.4 Internal intensity functions

Introduce the internal intensity function with three arguments (not to be confused with the internal intensity function with four arguments for the case of parallel illumination),

$$I(t, v, x) \quad = \quad \text{total intensity, at altitude } t \text{ in a slab of thickness } x, \text{ of energy going with direction cosine } v, \text{ and due to a distribution of sources } g(y),$$

$$0 \le t, y \le x; \; -1 \le v \le +1 \qquad (2.66)$$

We introduce the function

$$b(t, v, x) \quad = \quad \text{total intensity, at altitude } t \text{ in a slab of thickness } x, \text{ of energy going with direction cosine } v, \text{ and due to uniform isotropic illumination at the top,}$$

$$0 \le t \le x, \; -1 \le v \le +1,$$

$$= \quad 2 \int_0^1 I(t, v, x, u) du/u, \qquad (2.67)$$

where

$$I(t, v, x, u) \quad = \quad \text{internal intensity, at altitude } t \text{ in a slab of thickness } x, \text{ of energy going with direction cosine } v, \text{ and due to uniform parallel illumination at the top having direction cosine } u, \; 0 \le t \le x;$$

$$-1 \le v \le +1, \quad 0 \le u. \qquad (2.68)$$

The invariant imbedding equation,

$$I(t, v, x + \Delta) = I(t, v, x) + u(x, x)\Delta b(t, v, x) + o(\Delta) \qquad (2.69)$$

leads to the Cauchy problem,

$$I_x(t, v, x) \quad = \quad \left\{ g(x) + [\lambda(x)/2] \int_0^1 e(v, x) dv \right\}$$

$$\cdot \left\{ (1/2) \int_0^1 I(t, v, x, u) du/u \right\}, \qquad (2.70)$$

$$t \le x \le x_1,$$

$$I(t, v, t) \quad = \quad e(t, v), \; 0 \le t. \qquad (2.71)$$

Observe that the equations for the internal intensity function with parallel illumination, $I(t, v, x, u)$ must be adjoined, as well as those for $e(v, x)$.

2.5.5 Analytical derivation of Cauchy problems

The above Cauchy problems may be analytically derived from the integral equation for the source function,

$$U(t,x) \;=\; g(t) + [\lambda(t)/2] \int_0^x E_1\left(|t-y|\right) U(y,x) dy,$$

$$0 \le t \le x \le x_1, \tag{2.72}$$

where $E_1(s)$ is the first exponential integral function,

$$E_1(s) = \int_0^1 e^{-s/v} \frac{dv}{v},$$

and the expression for the emergent intensity is

$$e(v,x) \;=\; v^{-1} \int_0^x exp(-(x-y)/v) U(y,x) dy,$$

$$0 \le v \le 1;\; 0 \le x \le x_1. \tag{2.73}$$

Please refer to Appendix B for the details.

2.6 Reflecting Surfaces

For those studying terrestrial radiation and planetary physics, e.g., the estimation of cloud heights or the determination of ozone levels [7], it is important to take into account the effects of the surface underlying the atmosphere of interest. This is all the more true when the atmosphere is optically thin. As we have seen, invariant imbedding has proven itself in the ease of computing radiation patterns for scattering atmospheres as atmospheric thickness increases, starting from zero thickness to a given value. We have observed interesting changes in these patterns. We can regard the previous treatments to be suitable for the case of a perfectly absorbing surface. We turn our attention now to the question: What does invariant imbedding say about the scattered radiation patterns, internal and external, and their dependence on optical thickness when there is an underlying reflecting surface? We shall see that the Cauchy problem is only slightly modified.

2.6.1 Lambert surface reflector

Please review the derivation of the invariant imbedding variational equation presented earlier in this chapter for the case in which the atmosphere lies over a perfect absorber. Then, by adding a thin slab to the top of the slab, repeat the derivation for the case of an isotropically reflecting surface. The same variational equation is obtained, namely, Eq. (2.18). The numerical values of this S function differ from the corresponding previous ones because the initial values are different, as shown immediately below. Although the derivatives are computed according to the same procedure, the *values* of the S function used on the right hand sides are *different* and greater .

We consider the initial condition on the diffuse reflection function when the surface is a Lambert surface. A Lambert surface reflects the fraction A of downwelling radiation back such that the intensity is a constant, c.

$$\text{The upward flux} = A \text{ times the downward flux.} \qquad (2.74)$$

When the slab is of thickness zero, i.e., when there is no atmosphere, then the downward flux at the surface is $\pi F u$, where u is the direction cosine of incident parallel rays. Therefore, the right hand side of the above equation is $A \pi F u$.

Now, the upward flux (energy per unit time per unit solid angle per unit horizontal area) is expressed as

$$2\pi F \int_0^1 cv\,dv = \pi cF. \qquad (2.75)$$

But, by definition of the S function, the intensity of the upwelling radiation is

$$c = S(0, v, u)/4v. \qquad (2.76)$$

From the above, it is clear that $c = Au$ and that the S function for zero thickness is given by

$$S(0, v, u) = 4Auv. \qquad (2.77)$$

This scattering function is symmetric in u and v, because the initial conditions and the variational equations are symmetric.

By carrying out an imbedding derivation for transmission, we see that the variation with respect to thickness is the same as before, Eq. (2.21), as is the initial condition, Eq. (2.22). Observe, however, that evaluation of the transmission function will differ because of the coupling with the scattering function for the slab having surface albedo A.

2.6.2 Computational method and results

The invariant imbedding computational method for obtaining the diffuse reflection and transmission functions for an inhomogeneous slab with an underlying Lambertian surface follows the same procedure as before (Section 2.3), although the initial conditions are not identically zero but are given by the previous equations. The computer program written for this case will specialize to the case with the absorber when A is put equal to zero.

In this example, the inhomogeneous medium has an albedo for single scattering that depends on optical altitude t according to the model,

$$\lambda(t) = 0.5 + 2.0t - 2.0t^2, \quad \text{for } 0 \leq t < 1.0. \tag{2.78}$$

The S function for $A = 0$ is shown in Fig. 2.27, and for $A = 0.5$ in Fig. 2.28. The horizontal axis is the cosine of the emergent polar angle, so that 90 degrees (grazing angle) is at the left and zero degrees (vertical) is at the right. Remember that to determine intensity, the S function must be divided by $4v$. The seven curves represent incident directions close to normal (top) through close to grazing (bottom). These results were obtained by using a seven-point Gaussian quadrature formula with an Adams-Moulton fourth-order integration method and a step size of 0.01 in optical thickness, thus requiring the integration of a system of 28 nonlinear differential equations (using the symmetry of S) and 100 steps [4].

The Fortran computer program for producing the values of the S function follows the procedure of Fig. 2.6, as described earlier.

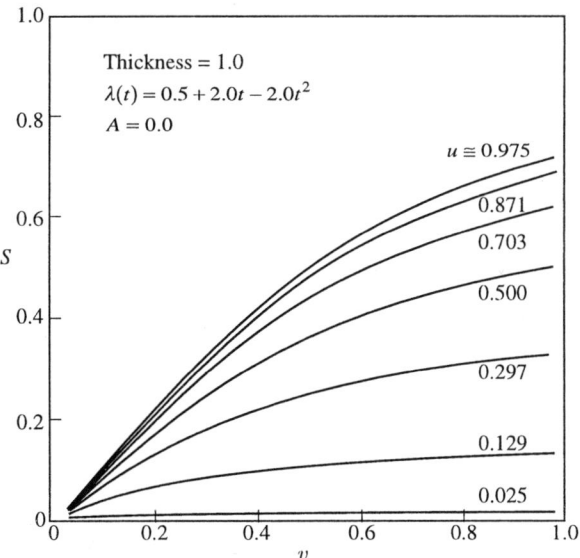

Figure 2.27. The S function for an inhomogeneous atmosphere with a completely absorbing bottom surface.

2.6.3 Specular reflector

Consider the case of a specular reflector characterized by the function $q(v)$, $0 \leq v \leq 1$, which represents the probability that a particle impinging on the

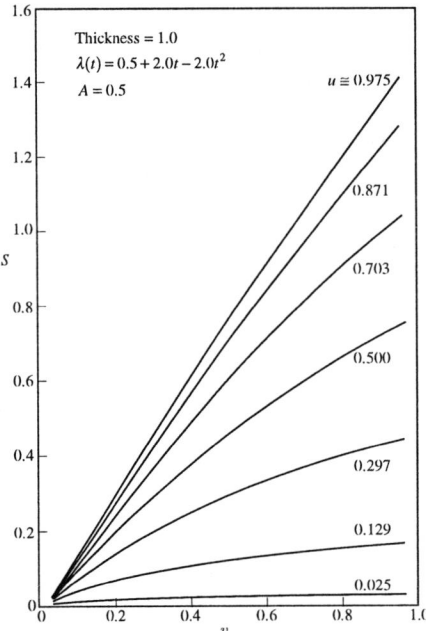

Figure 2.28. The S function for an inhomogeneous atmosphere with a Lambert surface having albedo 0.5.

bottom surface with angle of incidence $\arccos v$ is specularly reflected. The Cauchy problem for the S and T functions are given by the system,

$$
\begin{aligned}
S_x(v, u, x) &= -\left(\frac{1}{u} + \frac{1}{v}\right) S(v, u, x) \\
&\quad + \frac{\lambda}{4}\left[1 + q(u)e^{-2x/u} + 2\int_0^1 S(v', u, x)dv', v'\right] \quad (2.79) \\
&\quad \cdot \left[1 + q(v)e^{-2x/v} + 2\int_0^1 S(v, u', x)du', u'\right], \\
&\qquad x > 0,
\end{aligned}
$$

$$
S(v, u, 0) = 0 \qquad (2.80)
$$

$$
\begin{aligned}
T_x(v, u, x) &= -\frac{1}{u} T(v, u, x) \\
&\quad + \frac{\lambda}{4}\left[1 + q(u)e^{-2x/4} + 2\int_0^1 S(v', u, x)dv'/v'\right] \quad (2.81) \\
&\quad \cdot \left[e^{-x/v} + 2\int_0^1 T(v, u', x)du'/u'\right], \\
&\qquad x > 0,
\end{aligned}
$$

$$
T(v, u, 0) = 0. \qquad (2.82)
$$

The derivation is left to the reader. We simply comment that the exponential is due to the reduction of unscattered radiation along a slant path.

Tables and curves of computed values of reflected and transmitted intensities are reported in [4]. Source functions are treated in [18].

2.6.4 Equivalence relationships between cases with reflecting and absorbing surfaces

Lambert reflector. Oftentimes, one wants to know how much solar radiation reaches the surface of the earth. When we compare the fluxes for reflection and transmission from slabs with and without reflectors at the bottom, we discover some fascinating properties. Let us denote functions for the Lambert reflector with an asterisk; thus, the reflected intensity is $r^*(v, u, x, A)$, where A is the Lambert albedo, and the transmitted intensity is $t^*(v, u, x, A)$. The two cases are related as follows:

$$
\begin{aligned}
t^*(v, u, x, A) &= T(v, u, x)/4v \\
&\quad + \frac{I(u, x, A)}{\pi}\int_0^1 \frac{S(v, u', x)}{4x} 2\pi du', \quad (2.83) \\
&= T(v, u, x)/4v \\
&\quad + \frac{1}{2}I(u, x, A)\int_0^1 S(v, u', x)du', v,
\end{aligned}
$$

where the function $I(u, x, A)$ is the flux reflected from the bottom, given by

$$I(u, x, A) = \frac{A[ue^{-x/u} + \frac{1}{2}\int_0^1 T(v', u, x)dv']}{1 - A\int_0^1\int_0^1 S(v', u', x)du'dv'}. \tag{2.84}$$

Computational results in the form of tables and graphs confirm the above formulas. It is also observed, by comparison with the work of Kahle [11], that the reflected and transmitted fluxes due to isotropic scattering agree with those of Rayleigh scattering over albedoes $0 \leq A \leq 1$ and thicknesses up to 100. We have also observed that, for the conservative case of $\lambda = 1$ and $A = 1$, the reflected and transmitted fluxes are the same as those for the perfect specular reflector when the thickness is moderate, about three or greater.

Specular reflector. Let us turn our attention to the case of a homogeneous medium bounded from below by a perfect specular reflector, denoted by the asterisk. This case is compared with that of the perfect absorber. Consideration of the two source functions [12] leads to the relationship,

$$J^*(t, x, u) = J(x + t, 2x, u) + J(x - t, 2, u), \quad -x \leq t \leq x. \tag{2.85}$$

The interpretation of this equation is that, when the perfect specular reflector is removed, the medium is considered to be extended from thickness x to thickness $2x$ and illuminated from the top and bottom by parallel rays of net flux π incident at the original angle. There are other interesting equivalences for internal and external intensities in the same reference, such as

$$S^*(v, u, x) = S(v, u, 2x) + T(v, u, 2x), \tag{2.86}$$

and other more complex formulas involving the internal intensities b and h for omnidirectional illumination. Some of the abovementioned results have been confirmed by Kattawar [13] and extended to the case of inhomogeneous media with anisotropic scattering.

2.6.5 Discussion

These interesting results lead us to contemplate posing and solving inverse problems such as the following: Given measurements of the S function for a slab that can be modelled to have a parabolic albedo in t and an unknown Lambert surface albedo A, can we estimate the constants in the parabolic function as well as the surface albedo? Can we determine what type of reflection properties the surface has: Lambert, specular, etc.?

2.7 Omnidirectional Illumination

2.7.1 Introduction

In earlier parts of this book, we have been concerned with radiative transfer in atmospheres which are either illuminated by uniform parallel rays of radiation or those containing internal sources of radiation. When we carried out the invariant imbedding procedure in which a thin layer was added to the top of the slab, we saw that the thin layer contained scattering that contributed to internal and external intensities and to source functions. It was as if there were parallel rays of radiation incident on the slab from all directions, i.e., omnidirectional or isotropic illumination. So we used integrals of the corresponding functions to evaluate the effects internally and externally. This made it necessary to adjoin the equations for the parallel ray illumination to those for the emitting sources.

In this section we shall study the case of isotropic (omnidirectional) rather than monodirectional illumination of slabs and realize that this is the basic problem in radiative transfer in terms of which there are useful formulas and reductions. When there are emitting sources in an otherwise homogeneous atmosphere [14–16] (i.e., the albedo for single scattering is constant), then introduction of the b and h internal intensity functions of omnidirectional illumination reduces the number of independent variables in the invariant imbedding computation. The reason is that b and h functions do not depend on the incident direction.

For the case of monodirectional illumination of homogeneous slabs, the reduction is even more dramatic: all of the functions of interest (source, internal and external intensities) can be expressed *algebraically* in terms of b and h functions. Even for stratified media, the b and h functions are an interesting concept and merit further study.

The plan of this section is to introduce the b and h functions, to derive their invariant imbedding equations, to compute their solutions, and then to reexamine the monodirectional illumination and emitting source cases in terms of b and h functions.

2.7.2 The b and h functions

The functions $b = b(t, v, x)$ and $h = h(t, v, x)$ are *total* intensities defined as follows:

$b(t, v, x)$ = the total intensity at the altitude t in a direction whose direction cosine with respect to the upward normal is v, the slab thickness being x, and due to incident radiation at the *top* in all downward directions of one unit of energy per unit of horizontal area per unit of solid

angle per unit of time, (2.87)

and

$h(t, v, x)$ = the total intensity at the altitude t in a direction whose
direction cosine with respect to the upward normal is v,
the slab thickness being x, and due to incident radiation
at the *bottom* in all downward directions of one unit
of energy per unit of horizontal area per unit of solid
angle per unit of time, (2.88)

for $-1 \leq v \leq +1$.

When the slab is assumed to be homogeneous, we see that only "half" the functions b and h are required, since they are physically related,

$$b(t, v, x) = h(x - t, -v, x) \quad \text{and} \quad h(t, v, x) = b(x - t, -v, x). \quad (2.89)$$

This property simplifies the analysis and computation, for it is desirable to compute only the downwelling functions and determine the upwelling ones by the above formulas. For the trivial case of zero atmospheric thickness, the initial conditions which we'll need for invariant imbedding are readily seen to be, when $x = t$, for v being negative, i.e., $-1 \leq v < 0$,

$$h(t, v, t) = 0 \quad \text{and} \quad b(t, v, t) = |v|^{-1}. \quad (2.90)$$

We perform the imbedding as follows. A thin layer of thickness Δ is added to the top of the slab. For the function h, where v is negative, i.e., a downwelling direction, this results in the equation

$$h(t, v, x + \Delta) = h(t, v, x) + \left[\frac{\lambda}{4\pi} \int_0^1 Y(u', x) 2\pi du'/u' \right] \Delta b(t, v, x) + o(\Delta),$$

(2.91)

where the coefficient of b in the equation is the rate of production of scattered radiation at the bottom, and we have replaced it by the expression,

$$\frac{\lambda}{4\pi} \int_0^1 Y(u', x) 2\pi du'/u'.$$

These scatterers acting like omnidirectional illumination on the slab of thickness x create the internal intensity b at altitude t. In the limit as $\Delta \to 0$, this equation becomes

$$h_x(t, v, x) = \frac{1}{2} \lambda b(t, v, x) \int_0^1 Y(u', x) du'/u'. \quad (2.92)$$

For the function b we add a slab of thickness Δ to the bottom of the slab, "flip over the slab," and simultaneously lower the point of observation by a distance delta, thus keeping the point at the same altitude t. The increase in b is due to having omnidirectional scatters "at the opposite side" of the slab, which create the internal intensity h at altitude t. The decrease in b is due to the change in it as it traverses from altitude $t + \Delta$ to t, when downwelling. This change is readily obtained from the transfer equation in Appendix A, namely,

$$vb_t(t, v, x) = -b(t, v, x) + \Phi(t, x), \tag{2.93}$$

where $\Phi = \Phi(t, x)$ is the source function due to omnidirectional illumination at the top. The Φ function is the rate of scattering of the radiation as expressed by the internal intensity in all directions,

$$\Phi(t, x) = \frac{\lambda}{2} \int_{-1}^{+1} b(t, v, x)dv.$$

The Φ function is given in terms of the source function for monodirectional illumination by

$$\Phi(t, x) = 2 \int_0^1 J(t, x, u')du'/u'. \tag{2.94}$$

The resultant differential-integral equation for b is

$$b_x(t, v, x) = \frac{1}{2}\lambda h(t, v, x) \int_0^1 Y(u', x)du'/u' + v^{-1}[b(t, v, x) - \Phi(t, x)]. \tag{2.95}$$

The rate of production of scattered radiation at the top can be seen from earlier discussions to be related to the function X of Eq. (2.55) as follows:

$$J(x, x, u) = \frac{\lambda}{4}X(u, x).$$

Also, the differential-integral equation system for the X and Y functions can be derived and found to be given by

$$X_x(u, x) = \frac{1}{2}\lambda Y(u, x) \int_0^1 Y(u', x)du'/u', \tag{2.96}$$

$$Y_x(u, x) = -u^{-1}Y(u, x) + \frac{1}{2}\lambda X(u, x) \int_0^1 Y(u', x)du'/u', \tag{2.97}$$

and the equation for J is

$$J_x(t, x, u) = -u^{-1}J(t, x, u) + \frac{1}{2}\lambda X(u, x) \int_0^1 J(t, x, u')du'/u'. \tag{2.98}$$

The initial conditions are, at $x = 0$,

$$X(u, 0) = 1 \quad \text{and} \quad Y(u, 0) = 1, \tag{2.99}$$

and, at $x = t$,

$$J(t, t, u) = (\lambda/4)X(u, t). \tag{2.100}$$

Thus, the complete system is given by eqs. (2.92), (2.94), (2.96) through (2.102). To obtain b (or h) in the horizontal direction, use the relation,

$$b(t, 0, x) = \Phi(t, x), \tag{2.101}$$

easily seen from the equation of transfer,

$$v b_t = -b + \Phi,$$

and setting $v = 0$.

2.7.3 Computational method and results

The computational method is analogous to that for computing the source function. All functions are discretized at the N points of the Gaussian quadrature formula. Bear in mind that we wish to compute the b and h functions in downwelling directions $v = -v'$, where $0 \leq v' \leq 1$ and v' takes on the N values of the roots of the shifted Legendre polynomials.

Computational results have been tabulated and plotted. Extensive tables are found in the Rand report [15]. Fig. 2.29 is a plot of the b function propagating in three directions (horizontal, 13 degrees up, and 13 degrees down, as a function of depth, for slabs of albedo 0.9 and thicknesses 5 and 10. This graph was plotted from the results of computing b at altitudes $0.0, 0.1, 0.2, \ldots$, as the thickness was increased from zero up to ten. The b function is thus known numerically for all thicknesses between zero and ten. Observe how these intensities have a different behavior than the monodirectional intensities.

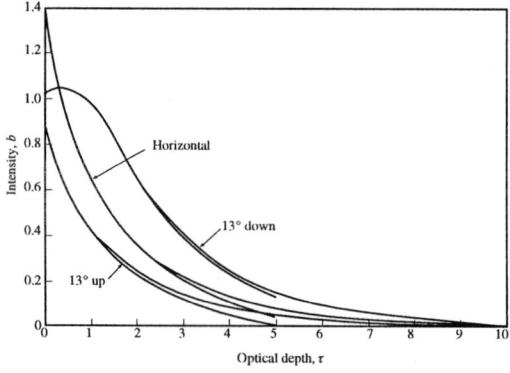

Figure 2.29. Intensities in three directions due to an isotropic (omnidirectional) input at the top, for slabs with albedo 0.9 and thicknesses 5 and 10.

2.7.4 Monodirectional illumination

The omnidirectional and monodirectional problems may be considered to be equivalent in the following way [17]. The omnidirectional functions such as $b(t, v, x)$, $h(t, v, x)$, $\phi(t, x)$ are expressible as integrals of the corresponding monodirectional functions $I(t, v, x, u)$ and $J(t, x, u)$, for homogeneous or inhomogeneous slabs, e.g.,

$$b(t, v, x) = 2 \int_0^1 I(t, v, x, u')du'/u' \qquad (2.102)$$

The converse, for homogeneous slabs, is not so obvious: The monodirectional functions may be expressed as algebraic functions of the b and h functions.

From the meanings of the X and Y functions as related to rates of production of scattered radiation at the top and the bottom, respectively, we have

$$X(u, x) = 1 + ub(x, u, x), \qquad (2.103)$$

$$Y(u, x) = uh(x, u, x), \qquad (2.104)$$

and clearly, from the definition of the source function $\Phi(t, x)$ and the equation of transfer in the horizontal direction,

$$\Phi(t, x) = b(t, 0, x). \qquad (2.105)$$

In order to derive a decomposition formula for the internal intensity with monodirectional illumination, namely $I = I(t, v, x, u)$, we have the transport equation when we "follow a beam,"

$$-vI_t = -I + J. \qquad (2.106)$$

On the other hand, we can compare the intensities at two nearby altitudes t and $t + \delta$ by removing a slab of thickness δ from the top and adding it to the bottom. Temporarily omitting all arguments except for the altitude, we have the finite difference equation

$$I(t+\Delta) - I(t) = \Delta u^{-1}I(t) - J(x, x, u)\Delta b(t, v, x) + J(0, x, u)\Delta h(t, v, x) + o(\Delta). \qquad (2.107)$$

In the limit as Δ tends to zero, this equation becomes

$$I = +u^{-1}I(t) - J(x, x, u)b(t, v, x) + J(0, x, u)h(t, v, x). \qquad (2.108)$$

Now, equating the two expressions for I from Eqs. (2.93) and (2.98), we obtain the desired formula,

$$(u^{-1} + v^{-1})I(t, v, x, u) = v^{-1}J(t, x, u) + J(x, x, xu)b(t, v, x) - J(0, x, u)h(t, v, x), \qquad (2.109)$$

or, in terms of the X and Y functions,

$$(u^{-1} + v^{-1})I(t, v, x, u) = v^{-1}J(t, x, u) + (\lambda/4)X(u, x)b(t, v,)$$
$$-(\lambda/4)Y(u, x)h(t, v, x). \qquad (2.110)$$

The decomposition formula for the source function is readily obtained by recognizing that J is the internal intensity in the horizontal direction and setting $v = 0$ in the last equation:

$$J(t, x, u) = (\lambda u/4)[X(u, x)b(t, -u, x) - Y(u, x)h(t, -u, x)] \qquad (2.111)$$

where X and Y are to be replaced by the right hand sides of Eqs. (2.91) and (2.92). To summarize, these algebraic formulas readily give the numerical values of the X, Y, J, I, r, t, and Φ functions of radiative transfer in terms of b and h functions computed via invariant imbedding and tabulated for easy reference. The physical interpretations are sometimes direct, as in the case of X and Y functions, but less intuitive yet interesting in the case of the internal intensity function $I(t, v, x, u)$. To this date, we know of no one-step derivation of the basic decomposition formula.

2.7.5 Internal emitting sources

We discussed in Section 2.5 the invariant imbedding treatment for the internal and external intensities for the case of an internal distribution of internal sources, $B(t)$. In terms of the b and h functions, the system of equations for the internal intensity function $I(t, v, x)$ and the emergent intensity function $e(v, x)$ becomes (see Eqs. (8)–(14) in [16])

$$I_x = b(t, v, x)\left[B(x) + \frac{\lambda}{2}\int_0^1 e(v'x)dv'\right], \qquad (2.112)$$

$$e_x = -u^{-1}e + B(x)u^{-1}X(u, x) + \frac{\lambda}{2}X(u, x)u\int_0^1 e(v', x)dv', \qquad (2.113)$$

with initial conditions at $x = 0$,

$$e(v, 0) = 0, \qquad (2.114)$$

and, at $x = t$,

$$I(t, v, t) = 0, \qquad (2.115)$$

and where the Cauchy problem for b, h, X, Y, and J has been presented above. The source function for emitting sources, $J^*(t, x)$, satisfies the equations

$$J_x^* = 2\left[B(x) + \frac{\lambda}{2}\int_0^1 e(v', x)dv'\right]\int_0^1 J(t, x, u')du'/u', \qquad (2.116)$$

and the initial condition at $x = t$,

$$J^*(t) = B(t) + \frac{\lambda}{2}\int_0^1 e(v', x)dv'. \qquad (2.117)$$

2.8 Discussion

In this chapter, we introduced the concept of invariant imbedding from the physical point of view, and we proceeded to apply it to derive the fundamental equations of reflection, transmission, internal intensities, and source functions. The source of radiation was either uniform parallel rays incident at the top of a slab or distributed sources of radiation within the slab. The computational method has been described and applied: first, for the basic equations for reflection and transmission; then for the adjoinment of a set of equations at each internal altitude where values of internal intensity and source functions are desired [18, 19]. A small sample of the abundant numerical results thus obtained has been presented. A partial picture of the complex behaviors of these interesting physical functions can be gleaned from this sampling and perhaps will serve to whet the reader's appetite for further computational investigations. But now, we turn to the inverse question: What physical system has given rise to these observed radiation patterns? On to the next chapter.

References

1. M. Abramowitz and I. Stegun (eds.), *Handbook of Mathematical Functions*, National Bureau of Standards, 1964.
2. R. Bellman, R. Kalaba and M. Prestrud, *Invariant Imbedding and Radiative Transfer in Slabs of Finite Thickness*, American Elsevier Publishing Co., New York, 1963.
3. R. Bellman, H. Kagiwada, R. Kalaba and M. C. Prestrud, *Invariant Imbedding and Time-Dependent Transport Processes*, American Elsevier Publishing Co., New York, 1964.
4. H. Kagiwada, R. Kalaba and S. Ueno, *Multiple Scattering Processes: Inverse and Direct*, Addison-Wesley Publishing Co., Reading, Mass., 1975.
5. B. Carnahan, H. Luther and J. Wilkes, *Applied Numerical Methods*, Wiley, New York, 1969.
6. W. Press, et al., *Numerical Recipes in C*, Cambridge University Press, Cambridge, 1992.
7. J. V. Dave, *J. Quant. Spect. Rad. Transfer*, Vol. 8, 1968, p. 25.
8. R. Bellman, H. Kagiwada, R. Kalaba and S. Ueno, "A Computational Approach to Chandrasekhar's Planetary Problem," *J. Franklin Institute*, Vol. 282, 1966, pp. 330–334.
9. J. Casti, R. Kalaba, and S. Ueno, "Reflection and Transmission Function for Finite Isotropically Scattering Atmospheres with Specular Reflectors," *J. Quant. Spectr. Rad. Transfer*, Vol. 9, 1969, pp. 537–552.
10. J. Casti, H. Kagiwada, and R. Kalaba, "External Radiation Fields for Isotropically Scattering Finite Atmospheres Bounded by a Lambert Law Reflector," *J. Quant. Spect. Rad. Transfer*, Vol. 10, 1970, pp. 637–651.
11. A. B. Kahle, "Global Radiation Emerging from a Rayleigh Atmosphere of Large Optical Thickness," *Astrophys. J.*, Vol. 151, 1968, pp. 637–645.
12. J. Casti, H. Kagiwada and R. Kalaba, "Equivalence Relationships Between Diffuse Radiation Fields for Finite Slabs Bounded by a Perfect Specular Reflector and a Perfect Absorber," *J. Quant. Spectr. Rad. Transfer*, Vol. 13, 1973, pp. 267–272.
13. G. W. Kattawar, "Solutions of the Equation of Transfer for a Medium Bounded by a Perfect Specular Reflector in Terms of Those for a Perfect Absorber," *J. Quant. Spect. Rad. Transfer*, Vol. 14, 1974, pp. 157–158.
14. H. Kagiwada and R. Kalaba, "A New Initial-Value Method for Internal Intensities in Radiative Transfer," *Astrophys. Journal*, Vol. 147, 1967, pp. 301–309.
15. H. Kagiwada and R. Kalaba, "Numerical Results for Internal Intensities in Atmospheres Illuminated by Isotropic Sources," *The Rand Corporation*, RM-4958-PR, 1966.
16. H. Kagiwada and R. Kalaba, "Initial-Value Methods for the Basic Boundary-Value Problem and Integral Equation of Radiative Transfer," *J. Comput. Physics*, Vol. 1, 1967, pp. 322–329.

17.   H. Kagiwada and R. Kalaba, "The Equivalence of the Isotropic and Monodirectional Source Problems," *J. Quant. Spectrosc. Radiat. Transfer*, Vol. 8, 1968, pp. 843–846.

18.   R. Bellman, H. Kagiwada, R. Kalaba, and S. Ueno, "Invarinat Imbedding and the Computation of Internal Fields for Transport Processes," *J. Math. Anal. Appl.*, Vol. 12, 1965, pp. 541–548.

19. R. Bellman, H. Kagiwada and R. Kalaba, "Invariant Imbedding and a Reformulation of the Internal Intensity Problem in Radiative Transfer Theory," *Monthly Notices Royal Astronomical Soc.*, Vol. 132, 1966, 183–191.

3. Inverse Problems

In addition to model building, as we did in the earlier chapters, we care very much about fitting observations, i.e., indirect measurements, to models. This is at the core of remote sensing, temperature retrieval, prospecting for oil, medical diagnosis, and other inverse problems. In this chapter we present several methods for the systematic formulation of inverse problems, and we give explicit computational procedures for carrying them out. We study their effectiveness and their robustness with respect to errors in observations, in models, in initial estimates. We present results of numerous computational experiments. Studies such as these serve in the planning of experiments. The benefits of analysis in the planning stages of a project can help to avoid unfruitful experiments and inferior designs.

3.1 Introduction

The subject of remote sensing is intimately connected with inverse problems, i.e., estimation problems. Scientists, engineers, physicians and others have been motivated by the desire to understand what they have been observing: estimation of orbital parameters based on observed motion of heavenly bodies, locating the atoms in a crystal structure based on diffraction patterns, determining the potential of a particle scattering system based on experimental scattering data, performing medical diagnoses, prospecting for oil fields. These estimation problems are also known by engineers as system identification problems [1], [2] and fall clearly into the class of optimization problems.

The term, *inverse problems*, arises in contrast to *direct problems*, which are generally the more tractable of the two types. In the direct problems of Chapter 2, we assumed that we knew the local physical parameters of the multiple scattering problem and so we could predict the diffuse radiation field through mathematical modelling, formulation as Cauchy problem, and computational solution. In this chapter we tackle the more difficult inverse problem. We desire to use systematic methods to form our estimates of unknown aspects of the scattering systems which we observe through remote, i.e., nondirect or nonlocal, measurements.

Obviously, one systematic method of estimating N constants in equations is to assume, let's say, K values of each constant and consider the K^N direct problems, seeking the one with the values of the N parameters that best explains the observations. This is not at all practical when K and N are in the order of 10 and the problem is as complex as the type we treated in Chapter 2. Nor is it useful for problems of moderate size; yet inverse problems occur widely.

During the late 1950's and early 1960's, methods such as dynamic programming [3] and quasilinearization [4] were originated and developed to avoid such enormous amounts of trial and error computing. While working in the field of control theory and optimization, we realized that the newly emergent system identification techniques could apply to inverse problems in radiative transfer. This approach fitted in perfectly with the invariant imbedding formulations that we had already begun to establish for the direct problems of multiple scattering.

We first used quasilinearization in the inverse problem of estimating orbital parameters for a heavenly body [5]. Then we applied the method to the invariant imbedding formulation of diffuse reflection in a slab illuminated by uniform parallel rays of radiation [6]. This was also highly successful. There were, however, two drawbacks: (1) the evaluation of partial derivatives of complicated nonlinear expressions, and (2) finding initial estimates of the unknown parameters. Fortunately, at the time of this writing, these difficulties have been completely eliminated. The first was resolved by the computational technique which is known as FEED [13] and the second by the method of associated memories. Now inverse problems can be treated in a practical, systematic manner. This is made possible in part due to invariant imbedding. Now, let us proceed with the solution of inverse problems.

3.2 Associative memories

Associative memory neural networks [7] were designed to aid in pattern recognition and more recently have been studied by Kalaba et al. in parameter estimation and inverse problems [8]. Associative memories have been used successfully in wide-ranging types of inverse problems: navigation (passive ranging and location), mechanical engineering, urban planning (attractiveness model of shopping centers), and in radiative transfer. Since associative memories are a type of neural network, they must first be trained with paired sets of measurement and parameter vectors. The training information is summarized in an associative memory matrix. Then the associative memory matrix is utilized to estimate a parameter vector when a measurement vector becomes available.

In the radiative transfer problems described below, we take as simulated data the reflected intensities presented in the tables of Chapter 2. We hold the incident angle constant at the seventh angle, 13.0 degrees. The reflected intensity observations used in the numerical experiments reported herein are the seven entries in the seventh row. As we know, these values have been very carefully computed. When noisy observations are wanted for numerical experiments, we round these values to the desired number of significant places.

We are interested in such questions as:

How well can albedo and thickness be estimated in controlled experiments in which the unknown parameters are actually found in the training sets?

What training sets lead to good estimates of parameters? How many pairs of training vectors, and which ones?

How does accuracy in the observations affect accuracy of the estimates?

Can estimates be extrapolated, i.e. found, even if they lie outside the training region?

We also want to understand how and why the associative memory method produces such good results in this application which is based on a nonlinear system of integro-differential equations. The answer to this last question is not yet at hand, although some intuition may be gained by careful perusal of the graphs of intensity patterns for different albedoes, thicknesses, and directions.

Investigation in progress is aimed at determining alternative memory matrices and selecting the "best" memory matrix to use, the "best" algorithm for computing it, rules of thumb for knowing in advance which training sets to use, how much error in observations can be tolerated in the basic method, and what to do when errors exceed this level (see, e.g., [9]).

3.2.1 Associative memory method

The associative memory estimator is written in the form,

$$a = M s \tag{3.1}$$

where M is the associative memory matrix, s is the measurement vector (column vector) and a is a parameter vector estimate (column vector). The tasks are to: (1) create a pair of training matrices from given measurements and parameters, (2) form the memory matrix M from the training data, and (3) employ the memory matrix to estimate an unknown parameter vector from a given measurement vector.

In stage 1, the measurement vectors are gathered into a rectangular matrix, S. The corresponding known parameter vectors are likewise gathered into a rectangular matrix, A. The columns of the S and A matrices are the respective pairs of training vectors. While the actual ordering of the pairs is immaterial, they must be in one-to-one correspondence; i.e., the first column of the S matrix must associate with the first column of the A matrix, and so on. There are as many columns in S and A as there are training cases. The number of rows in S is the number of observations in each training case. The number of rows in A is, of course, the number of components of the parameter vector. For example, if we are seeking to estimate albedo only, then there would be one component and one row; if we are estimating albedo and thickness, then there would be two components and thus two rows. We have the approximate relation,

$$A = MS, \tag{3.2}$$

where M is as yet unknown. Since both A and S are known, M can be computed.

This is done in Stage 2. We choose to determine M as the matrix that satisfies Eq. (3.2) in a least squares sense. Then M is expressed as

$$M = AS^+, \tag{3.3}$$

where S^+ is the Moore-Penrose generalized inverse of the matrix S. Matrix M has as many rows as there are rows in the parameter vector and as many columns as there are observations in the measurement vector.

In Stage 3, the memory matrix is used to estimate the unknown parameter vector a when an observation vector s is given:

$$a^* = Ms \tag{3.4}$$

3.2.2 Computational procedure

The general computational procedure for estimating parameters of diffusely scattering systems based on reflected radiation measurements is shown in Table 3.1.

Table 3.1. Computational Procedure for Associative Memory Estimation

1. Create training sets
a) Input measurement vectors into matrix S
b) Input corresponding parameter vectors into matrix A
2. Compute memory matrix
a) Compute S^+
b) Compute $M = AS^+$
3. Compute estimate of parameter vector
a) Input measurement vector s
b) Compute estimate $a^* = Ms$

3.2.3 Computational experiments

Controlled experiments demonstrate the effectiveness of this method. We use simulated measurement data which are obtained by solving the invariant imbedding equations of Chapter 2 with known parameter values. After selecting a set of training matrices, we calculate the memory matrix and employ it to form estimates. We find that the method produces acceptable and even quite accurate values, even though the estimator is linear and the data come from a nonlinear system. Experimentation with different training sets is advisable prior to use of the memory matrix. The process is quite robust, and only a few training sets are required.

In order to evaluate the accuracy of the estimation, we compute A^*, the matrix of estimated parameter vectors, for each member of the original training set using the equation,

$$A^* = MS \tag{3.5}$$

where S is the measurement matrix of the training set. Then A^* is compared with A.

This computational method can be implemented in Fortran, in Matlab, Mathcad, or other language or software.

3.2.4 Albedo estimation

Stage 1. We train the associative memory with ten cases having albedoes of $0.1, 0.2, \ldots, 1.0$, for which seven noisy reflected intensities are obtained for the case of illumination at 13 degrees. The noise-free values of intensities are obtained through computing the solution of the Cauchy problem of the preceding sections. Random noise [10] in the amount of 5% is added to simulate intensity measurement errors. The A and S training matrices have seven rows and ten columns.

Stage 2. The memory matrix has dimension one row and seven columns. We present the memory matrix obtained, rounded to one decimal place (of course, when used, the elements in the matrix must be known quite precisely):

$$M = (7.5 \quad -3.8 \quad -0.9 \quad 2.6 \quad 2.1 \quad 3.5 \quad -9.3) \tag{3.6}$$

Stage 3. For each albedo, we produce new noisy observations with 1% noise, and we repeat this 50 times each (a total of 500 Monte Carlo runs). The following are the mean estimates when the true albedoes are $0.1, 0.2, \ldots, 1.0$:

$$0.09 \quad 0.18 \quad 0.27 \quad 0.37 \quad 0.47 \quad 0.58 \quad 0.70 \quad 0.82 \quad 0.94 \quad 1.03 \tag{3.7}$$

These results are accurate to 10% or better. Of course, when we add too much noise to the measurements, such as 10%, the parameter estimates become quite poor. However, the estimates can be sharpened even with this large amount of noise using a nonlinear modification suggested by Poggio [9]. The above results, obtained via a Matlab program on a Macintosh computer, were first reported in [8] and were confirmed with a Fortran program on a VAX computer.

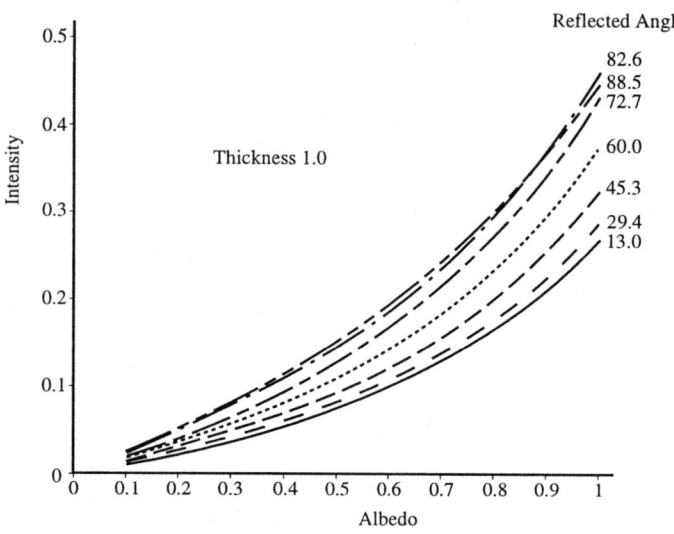

Figure 3.1. Reflected Intensities in Different Directions for Multiple Scattering in Slabs with Thickness 1.0 as Functions of Albedo.

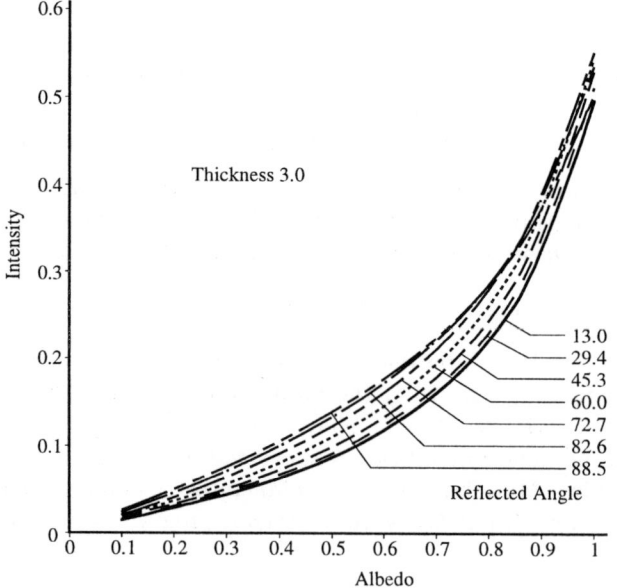

Figure 3.2. Reflected Intensities in Different Directions for Multiple Scattering in Slabs with Thickness 3.0 as Functions of Albedo.

Table 3.2. Estimation of Thickness for Various Albedoes

Albedo	Thicknesses	Digits in Input	Estimates of thickness
0.1	1.0 2.0 3.0	6	0.95 2.4 2.6
		5	0.95 2.4 2.6
		4	0.96 2.5 3.0
		3	1.00 2.0 3.0 (to 9 digits)
0.5	1.0 2.0 3.0	6	1.0(9) 2.0 3.0
		5	1.0(8) 2.0 3.0
		4	1.0(6) 2.0 3.0
		3	1.0(7) 2.0 3.0
0.9	1.0 2.0 3.0 5.0	6	1.00 2.00 3.00 5.00
		5	1.05 2.06 3.06 5.05
		4	1.03 2.03 3.03 5.03
		3	1.003 2.003 3.003 5.002
1.0	1.0 2.0 3.0	6	1.0(9) 2.0 3.0
		5	1.0(9) 2.0 3.0
		4	1.0(9) 2.0 3.0
		3	1.0(9) 2.0 3.0

3.2.5 Estimation of thickness

The thickness of each of 13 atmospheres was estimated in a series of experiments. Some of the training cases are shown in Fig. 3.1 for thickness 1.0 in a plot of reflected intensity in seven directions for ten slabs of albedoes 0.1, 0.2,...,1.0. Fig. 3.2 is a similar graph for thickness 3.0. Comparison of these two figures shows differences in the relationships among the curves. Results for these 13 cases are presented in Table 3.2. True values of thickness for four training cases are given in the second column; the estimates in column four are to be compared against these, respectively. The number of digits in the input data is varied from six to three in order to study the effect of observational accuracy. A number in parentheses indicates the number of correct zeros after the decimal point. No serious degradation of estimates is found here due to decreased number of correct digits in the data. It may seem a surprise that thickness estimates for albedo 0.1 were poorer than the others, but observe that the intensity values are all very small.

3.2.6 Estimation of thickness with different training sets

Experiments concern the case of conservative scattering, albedo 1.0, which displays properties markedly different from those cases for which albedo is less than 1.0. The intensities used in the training sets are shown in Figs. 3.3 and 3.4. Fig. 3.3 shows intensities plotted against thickness for the seven angles of reflection, the seven components of the observation vectors for those five thicknesses. Fig. 3.4 shows a family of curves for five different thicknesses plotted against angle of reflection. Table 3.3 shows six sets of training cases, with three, four, or five thicknesses. Observe that the poorest results in Table 3.3 were obtained for five and four thicknesses in the training cases; the inclusion of thickness 10.0 served to "confuse" the memory matrix when several other thicknesses were involved. This is explained by the uppermost curve in Fig. 3.4 and the crossovers in Fig. 3.3. On the other hand, with only two other thicknesses, the inclusion of thickness 10.0 leads to satisfactory results. The other estimates shown are highly accurate.

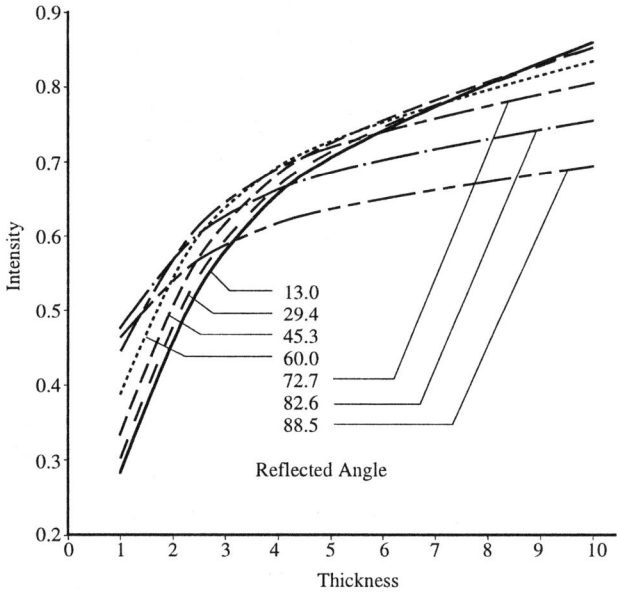

Figure 3.3. Reflected Intensities in Different Directions as Functions of Thickness.

3.2.7 Extrapolation and interpolation in estimation of thickness

Extrapolation and interpolation cases are examined in Table 3.4. The extrapolation to true thickness 5.0, estimated thickness 4.3, is shown in the first group of results using training cases with thickness values 1.0, 2.0, 3.0. The remainder of the experiments show interpolation estimates with varying accuracies, depending upon the choice of training cases as compared with the actual thickness. For example, the thickness 3.0 is estimated as 3.3 when thicknesses 1.0, 2.0 and 5.0 are used. With training thicknesses 1.0, 3.0 and 10.0, the thickness 2.0 is estimated to be approximately 0.8 and thickness 5.0 is estimated to be 6.4.

3.2.8 Percent error in estimates of thickness

Table 3.5 focuses on the percent error in estimates (true minus estimated values) for two sets of training/estimated cases; thicknesses 1.0, 2.0 and 3.0,

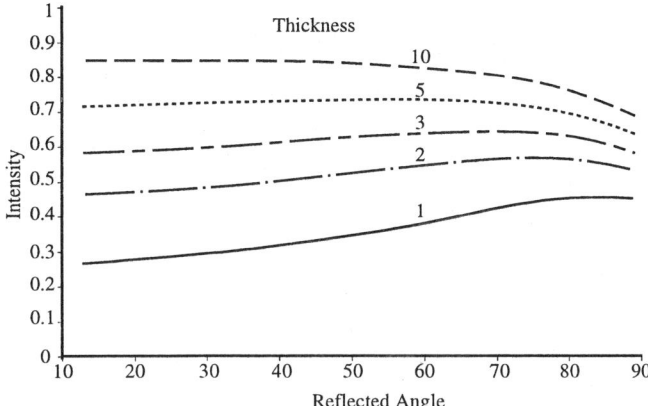

Figure 3.4. Reflected Intensities Versus Reflected Angle for Slabs of Various Thicknesses.

and 1.0, 3.0 and 10.0. Observe the excellent results, no matter what the true thickness, the training thicknesses, or the number of digits in the data.

3.2.9 Discussion

The results presented above indicate that the associative memory method is of value in obtaining estimates of unknown parameters in the considered system. The user of the method must select training cases judiciously and conduct an investigation over cases which are likely to be encountered. The estimates from associative memories may be used as preliminary estimates for refinement using the quasilinearization method.

The authors are indebted to B. K. Shams for assistance with the computational work reported in this section.

3.3 Quasilinearization

In the event that the initial estimates obtained through associative memories are not acceptable, a refinement process can be applied. Kalaba's method of quasilinearization is similar to Newton's method for nonlinear least squares, except that the model equations are differential equations [4]. These differential equations are generally nonlinear, either because they were nonlinear in the direct problem, or because the introduction of unknown parameters as other dependent variables makes even a linear equation nonlinear. Our model equations are the invariant imbedding equations of radiative transfer, and the parameter vector may be comprised of the albedo for single scattering, the

Table 3.3. Roundoff Effect on Estimates of Thickness for Albedo 1.0

Albedo	Thicknesses	Digits in Input	Estimates of thickness
1.0	1.0 2.0 3.0 5.0	6	1.001 2.002 3.002 5.002
		5	1.004 2.005 3.006 5.006
		4	1.004 2.005 3.006 5.006
		3	1.001 2.002 3.002 5.002
1.0	1.0 2.0 3.0 5.0 10.0	6	15.2 20.5 23.6 27.0 30.3
		5	14.5 20.1 23.5 27.2 30.8
		4	14.3 20.1 23.5 27.2 30.9
		3	14.3 20.1 23.5 27.2 30.9
1.0	3.0 5.0 10.0	6	3.0(7) 5.0 10.0
		5	3.0(6) 5.0 10.0
		4	3.0(7) 5.0 10.0
		3	3.0(7) 5.0 10.0
1.0	2.0 3.0 5.0 10.0	6	1.71 2.64 4.59 9.60
		5	4.17 5.44 7.61 12.32
		4	4.46 5.78 8.00 12.69
		3	2.24 3.27 5.29 10.26
1.0	1.0 3.0 10.0	6	1.0(9) 3.0(9) 10.0(9)
		5	1.0(9) 3.0(9) 10.0(9)
		4	1.0(9) 3.0(9) 10.0(9)
		3	1.0(9) 3.0(9) 10.0(9)
1.0	1.0 2.0 5.0	6	1.0(9) 2.0(10) 5.0(10)
		5	1.0(9) 2.0(10) 5.0(10)
		4	1.0(9) 2.0(10) 5.0(10)
		3	1.0(9) 2.0(10) 5.0(10)

coefficient of the surface albedo, the coefficients in the expansion of the phase function, and others.

3.3.1 Model equations

Consider a physical system which is described by the system of N equations

$$dx/dt = f(x, a), \qquad 0 \le t \le T \tag{3.8}$$

where x is a vector of dimension N, a function of independent variable t, with the N initial conditions

$$x(0) = c. \tag{3.9}$$

The vector x describes the state of the system at "time" t, and a is a parameter vector of dimension $M < N$ of the system. With the value of a given, the state at any time t, $x(t)$, may be calculated by a numerical integration of (3.8) with initial conditions (3.9).

Table 3.4. Extrapolated and Interpolated Estimates of Thickness

Albedo	Thickness	Digits in Input	Estimate of Thickness	True Thickness
1.0	1.0 2.0 3.0	6	4.226	5.0
		5	4.226	5.0
		4	4.270	5.0
		3	4.262	5.0
1.0	1.0 2.0 5.0	6	3.290	3.0
		5	3.290	3.0
		4	3.288	3.0
		3	3.291	3.0
1.0	1.0 3.0 10.0	6	0.8214	2.0
		5	0.8213	2.0
		4	0.8294	2.0
		3	0.8347	2.0
1.0	1.0 3.0 10.0	6	6.3807	5.0
		5	6.3810	5.0
		4	6.3837	5.0
		3	6.3683	5.0

Table 3.5. Percent Error of Estimates of Thickness

Albedo	Thicknesses	Digits in Input	Percent Error		
1.0	1.0 2.0 3.0	6	0.228E-7	0.134E-7	0.877E-8
		5	0.264E-7	0.158E-7	0.105E-7
		4	0.355E-7	0.216E-7	0.146E-7
		3	0.184E-7	0.113E-7	0.784E-8
1.0	1.0 3.0 10.0	6	0.917E-8	0.443E-8	0.107E-8
		5	0.203E-7	0.998E-8	0.263E-8
		4	-0.159E-7	-0.808E-8	-0.239E-8
		3	0.687E-8	0.338E-8	0.101E-8

3.3.2 Inverse problem

Now let us suppose that we have a system described by Eqs. (3.8) and (3.9), but a is unknown to us. However, we are able to make measurements of the components of the state of the system at the end of the interval, T. We wish to identify the system by determining a.

We think of the system parameter vector as if it were a dependent variable which satisfies the vector equation, which represents M scalar equations,

$$da/dt = 0 \qquad (3.10)$$

with the unknown initial conditions

$$a(0) = e. \qquad (3.11)$$

The boundary value problem is to find the complete set of M initial conditions such that the solution of the nonlinear system of $N + M$ equations,

$$\begin{aligned} dx/dt &= f(x, a) \\ da/dt &= 0, \end{aligned} \qquad (3.12)$$

with initial conditions (3.9) and (3.11) agreeing with the approximate boundary conditions

$$x(T) = g \qquad (3.13)$$

where g is the observed state vector at time T, i.e., a measurement vector of dimension N. This is a case in which there are more measurements (N) than there are unknowns (M). We shall demonstrate the method with a least squares fit to the data, although clearly other criteria–such as minimax–could be used.

3.3.3 Quasilinearization problem

Quasilinearization is a method of successive approximations. A sequence of linear problems is solved. It is assumed that a high order numerical integration method is used that yields accurate solutions of large systems of nonlinear differential equations. It is further assumed in this book that the systems of linear algebraic equations encountered are accurately solved. In the event that the linear equations form an ill-conditioned system, then steps must be taken to glean an accurate solution, such as through Gram-Schmidt orthogonalization or dynamic programming.

Let us define a new column vector x of dimension $R = N + M$, having as its elements the N components of the original vector x and the M components of a, such that the last M components of x are

$$x_{N+j} = a_j, \qquad j = 1, 2, \ldots, M. \qquad (3.14)$$

This enlarged vector satisfies the system of R nonlinear equations

$$dx/dt = f(x), \qquad 0 \le t \le T, \qquad (3.15)$$

or, component-wise,

$$dx_i/dt = f_i(x), \quad i = 1, 2, \ldots, R, \qquad 0 \le t \le T, \qquad (3.16)$$

according to (3.5), and it has the N *known* initial conditions

$$x_i(0) = c_i, \qquad i = 1, 2, \ldots, N, \qquad (3.17)$$

where the c_i are the components of the vector c. It also satisfies the M *unknown* initial conditions

$$x_{N+j}(0) = a_j(0) = e_j, \qquad j = 1, 2, \ldots, M, \qquad (3.18)$$

and the approximate terminal conditions

$$x_i(T) \cong g_i, \qquad i = 1, 2, \ldots, N. \qquad (3.19)$$

Let us replace the conditions (3.19) by the least squares fit criterion,

$$\min \sum_{i=1}^{N} [x_i(T) - b_i]^2 \tag{3.20}$$

where the minimization is to be carried out over the choices of the vector e. This will result in a system of M equations for the M unknowns.

3.3.4 Quasilinearization theory

An initial approximation starts the calculations. We form an estimate of the unknown vector e, and beginning with the initial conditions in (3.17) and (3.18) we numerically integrate the system (3.15) or (3.16) to product $y(t)$, the current approximation to the solution $x(t)$ on the entire interval $t = 0$ to $t = T$. This solution $y(t)$ is stored in the computer's memory at a grid of points. Review Chapter 2 on the topic of numerical integration of a system of differential equations with initial conditions.

We now describe the quasilinearization theory, and then we will describe the quasilinearization numerical method. Let us examine the equations which determine the next approximation from the current approximation and the observations. We linearize the nonlinear equations (3.15), using Taylor series expansion and keeping only the linear terms. The linearized equations for the new approximation to be called $x(t)$ (not to be confused with the exact solution $x(t)$) are

$$dx/dt = f(y) + J(y)(x - y), \tag{3.21}$$

where the Jacobian matrix has elements

$$J_{ij} = \frac{\partial f_i}{\partial x_j}, \qquad i, j = 1, 2, \dots, R. \tag{3.22}$$

Since the new approximation is a solution of a system of linear differential equations, we know from general theory that it may be represented as the sum of a particular solution vector, $p(t)$, and a linear combination of M independent vector solutions of the homogeneous equations, $h^i(t)$, $i = 1, 2, \dots, M$,

$$x(t) = p(t) + \sum_{i=1}^{M} e_i h^i(t), \qquad 0 \le t \le T. \tag{3.23}$$

Through substitution of (3.23) in (3.21), we see that the function $p(t)$ is a particular solution that satisfies the differential equation

$$dp/dt = f(y) + J(y)(p - y), \qquad 0 \le t \le T. \tag{3.24}$$

For convenience we choose the initial conditions

$$p(0) = 0. \tag{3.25}$$

The M functions $h^i(t)$ are separate solutions of the M homogeneous systems

$$dh^i/dt = J(y)h^i, \quad i = 1, 2, \ldots, M, \qquad 0 \le t \le T, \qquad (3.26)$$

and we choose the M sets of initial conditions,

$h^i(0)$ is the unit vector with all of its components zero, except for the ith which is one, $i = 1, 2, \ldots, M$. $\qquad (3.27)$

The $h^i(0)$ form a linearly independent set. If the interval $(0, T)$ is sufficiently small, the functions $h^i(t)$ are also independent. The solutions $p(t)$, $h^i(t)$ are produced by numerical integration with the given initial conditions. There are $M + 1$ systems of differential equations, each with M equations, making a total of $M(M + 1)$ equations which are integrated at each stage of our calculations.

After the vector functions p and h^i have been found over the interval, we combine them so as to satisfy the terminal conditions (3.20) which represent the fitting of the observations. Suppose that we have exactly $N = M$ perfectly accurate observations. There would be M algebraic equations in M unknowns of the form

$$Ac = b, \qquad (3.28)$$

where

$$A_{ij} = h_i^j(T), \qquad (3.29)$$

$$c_j = e_j, \qquad (3.30)$$

$$b_i = x_i(t) - g_i. \qquad (3.31)$$

This linear algebraic equation system can be solved using a method such as Gaussian elimination.

When observational data are utilized, they will contain errors, and there will be more observations than there are unknowns. For a least squares fit, (see also section 3.5.6) the M by M matrix A is

$$A_{ij} = [h_i^j(T)]^T h_i^j(T), \qquad (3.32)$$

where the superscript T represents the transpose of the indicated matrix. The components of the M-dimensional column vector b are

$$b_j = [h_i^j(T)]^T [x_i(T) - g_i]. \qquad (3.33)$$

This system is solved in the same way for the values in the multiplier vector, e.

Having determined the multipliers, we now know a complete set of initial conditions for the approximation in this stage. Because of our choice of initial conditions for p and h^i, the initial values for last M components of the vector x are identical with the multipliers,

$$a_j(0) = e_j, \qquad j = 1, 2, \ldots, M. \tag{3.34}$$

Furthermore, we have a new approximation to the system parameter vector a.

The new approximation $x(t)$ for the interval $(0, T)$ may be produced either by the integration of the linear equations with the initial conditions just found, or by the linear combination of $p(t)$ and $h^i(t)$. This cycle is now complete, and we are ready for the next. The process may be repeated until no further change is noted in the vector e.

3.3.5 Quasilinearization method

The quasilinearization procedure is summarized in Table 3.6.

Table 3.6. Quasilinearization Numerical Method

---

1. Input data
   - a) Enter initial estimates of vector e
   - b) Enter observation vector b
2. Produce initial approximation
   - a) Set up initial conditions for nonlinear differential equation system
   - b) Integrate nonlinear differential equation system from 0 to T, storing at all grid points
3. Do iterations
   - a) Set up initial conditions for linear differential equation systems for particular and homogeneous solutions
   - b) Integrate respective systems from 0 to T, storing at all grid points
   - c) Set up matrix of coefficients and right hand vector for system of linear algebraic equations
   - d) Solve linear algebraic system, define the coefficients of the homogeneous solutions, and define the new estimates of the parameters
   - e) Form the new approximation by taking the linear combination of the particular and homogeneous solutions
   - f) Output the new estimates and the new approximation
   - g) Test for convergence: if converged, stop; otherwise continue

---

3.4 FEED automatic derivative evaluation

3.4.1 Introduction to FEED

The quasilinearization method has been extensively applied in inverse problems of radiative transfer. One might, at first, think that it would be excruciatingly cumbersome to form and evaluate the partial derivatives of the nonlinear right hand sides of the equations we have derived in above and in the following sections. Fortunately, this is not the case when automatic derivative evaluation is used. In implementing quasilinearization for a given problem, the Jacobian matrix of partial derivatives of the nonlinear functions with respect to each of the independent variables must be evaluated at each step of the integration in a given iteration. For all but the simplest functions, the forming of these partial derivatives "manually" is a tedious and error-prone chore. Since these partial derivatives are used in a computational scheme and are not of primary interest in themselves as analytical functions, we prefer to utilize an automatic derivative evaluation method. The one that we use is called FEED, for Fast and Efficient Evaluation of Derivatives.

FEED is a modification of a method introduced by R. Wengert [11] in 1963. First described in [12] and elaborated upon in the book [13], the FEED method is readily implemented in Fortran, C, C++, Ada, or other similar programming language. We illustrate the method using Fortran.

We consider the evaluation, at a certain "time," t, of the partial derivatives of the function

$$z = f(x, y), \tag{3.35}$$

where $x = x(t)$ and $y = y(t)$. Observe that "time" t corresponds to a specific step in our numerical integration. Derivatives of x, y, and z with respect to t are denoted with a dot, while partial derivatives of f with respect to x and y are denoted by subscripts, e.g., $\partial f / \partial x = f_x$. The chain rule of differential calculus,

$$\dot{z} = f_x \dot{x} + f_y \dot{y}, \tag{3.36}$$

is useful. Wengert pointed out that if we set

$$\dot{x} = 1 \text{ and } \dot{y} = 0, \tag{3.37}$$

then we would have from the chain rule that one of the desired partial derivatives at time t is

$$f_x(x, y) = \dot{z}. \tag{3.38}$$

In equation (3.38) all quantities are evaluated at the current "time." The question, then, is how does one evaluate the requisite total derivative? Before providing the answer, we observe that to obtain the other partial derivative, that with respect to y, we would set $\dot{y} = 1$ and $\dot{x} = 0$.

3.4.2 Description of FEED by an example

Consider the nonlinear function

$$z = \ln xy \tag{3.39}$$

where ln represents the natural logarithm function. We wish to evaluate the two partial derivatives of $z = f(x, y)$ at the current "time" when x has the value $x(t)$ and y has the value $y(t)$. Let us set up a table as shown below,

Table 3.7, in which new variable names are introduced in a sequential fashion, in such a way that when we arrive at the end of the table we will have the desired partial derivatives.

Table 3.7. Evaluation of the Partial Derivatives of $z = \ln xy$

Variable	$\partial/\partial x$	$\partial/\partial y$	Subroutine
$a = x$	$a_x = 1$	$a_y = 0$	lim 1
$b = y$	$b_x = 0$	$b_y = 1$	lim 2
$c = ab$	$c_x = a_x b + a b_x$	$c_y = a_y b + a b_y$	mult
$z = \ln c$	$z_x = c^{-1} c_x$	$z_y = c^{-1} c_y$	logg

The new variables in the table are dimensioned to be three, i.e., they are vectors with three components. The three components are the original variable itself (column one of the table), then its partial derivative with respect to x (column two), and then its partial derivative with respect to y (column three). The table is implemented in the following Fortran subroutine subprograms. The inputs are the current values of x and y and the output is the vector z. The first two rows are carried out through subroutines lin1 and lin2.

```
subroutine lin1(x, a)        subroutine lin2(y, b)
dimension a(3)               dimension b(3)
a(1) = x                     b(1) = y
a(2) = 1.0                   b(2) = 0.0
a(3) = 0.0                   b(3) = 1.0
return                       return
end                          end
```

The third row, for the forming of the product of x and y and of the product's partial derivatives, is done by subroutine mult(a, b, c).

```
subroutine mult(a, b, c)
dimension a(3), b(3), c(3)
c(1) = a(1)*b(1)
c(2) = a(2)*b(1) + a(1)*b(2)
c(3) = a(3)*b(1) + a(1)*b(3)
return
end
```

Subroutine logg(c, d) is written and executed to carry out the evaluation of the logarithm function and its partial derivatives, automatically, as shown in the fourth row of the table. The variable c is the argument of the logarithm, and the logarithm z and the indicated partial derivatives are computed and stored in the three components of the vector z.

```
subroutine logg(c, z)
dimension c(3), z(3)
z(1) = a log(c(1))
z(2) = c(2)/c(1)
z(3) = c(3)/c(1)
return
end
```

Finally, the execution of the procedure in the table is done through the subroutine that defines the nonlinear function z being considered,

```
subroutine fun(x, y, z)
dimension a(3), b(3), c(3), z(3)
call lin1(x, a)
call lin2(y, b)
call mult(a, b, c)
call logg(c, z)
return
end
```

To ensure that the reader has understood the implementation of **FEED**, the reader is advised to program the above subroutines together with a test main program that sets the desired values of x and y, calls subroutine fun, prints out the computed values of the function and the partial derivatives, and also prints out the true values computed from the evaluation of analytically derived partial derivatives for comparison's sake.

3.4.3 Remarks and extensions

Observe that there are four types of subroutines used by **FEED**: (1) vectorizations (or linearizations) of the scalars x and y into the a and b vectors; (2) operations such as multiplication (subroutine mult), addition, subtraction, division of functions, and the chain rule when needed (these are used as library functions); (3) functions such as the trigonometric, exponential and other functions encountered in the applications (these are also library functions), and (4) function evaluations such as subroutine fun for defining the actual nonlinear functions in one's application. To extend the FEED method to more general functions, to functions of more variables, and to higher order derivatives, one must write vectorization subroutines of appropriate dimension, and the set of library functions must be increased to include the needed ones. Once such subroutines have been written, then one only needs to write new function subroutines such as fun for the function evaluations of actual interest.

To incorporate FEED into quasilinearization, use the appropriate subroutines like fun to evaluate the partial derivatives of the Jacobian matrix of section 3.3.4.

3.5 Inverse problems for inhomogeneous media and effect of criterion on estimates

3.5.1 Inverse problems for layered media

Consider an inhomogeneous medium whose albedo for single scattering is described by the function

$$\lambda(t) = a + b\tanh 10(t - c), \qquad 0 \leq t \leq T \qquad (3.40)$$

where the constants a, b, c and T are to be estimated on the basis of 49 accurate reflected intensity measurements. The measurement data we use are those for the case,

$$a = 0.5, \quad b = 0.1, \quad c = 0.5, \quad T = 1.0. \qquad (3.41)$$

These true values serve as checks on the accuracy of the estimation. This medium has two distinct albedoes, $a - b$ in the lower half, $a + b$ in the upper half.

The reflected intensity function is a matrix that satisfies the Cauchy system of Chapter 2. Numerical results for the reflected intensities were presented in Chapter 2. The discretized $S_{ij}(x)$ for $i = 1, 2, \ldots, N$; $j = 1, 2, \ldots, N$, are organized into a column vector of dimension N^2. This column vector is identified as the state vector of section 3.2.

Using the quasilinearization method, we conduct a series of experiments to estimate the first constant c only, keeping the other constants fixed at their correct values. We use different initial estimates. In the first trial, the initial estimate is $c = 0.2$. The sequence of estimates via quasilinearization, shown in Table 3.8, demonstrates rapid, quadratic, convergence.

Table 3.8. Sequence of Quasilinearization Estimates of the Interface Constant, c

Approximation	Estimate of c
0	0.200000
1	0.620000
2	0.518700
3	0.500089
4	0.499990

Similar rapid convergence is obtained in the second trial, in which the initial estimate is $c = 0.8$. In a third trial, with $c = 0.0$ initially, there is no convergence. Obviously, this estimate is too far from the optimal choice of c.

3.5.2 Estimation of optical thickness

The estimation of optical thickness T poses the question of how to estimate the end of the interval of integration. We use the following "trick": we scale the interval of integration by introducing a new variable of integration, s, such that

$$st = T, \qquad 0 \leq s \leq 1. \tag{3.42}$$

The state vector becomes a function of the new independent variable, s. The state vector is extended in the inverse problem through the adjoining of a variable representing the unknown parameter T, and its differential equation is

$$dT/ds = 0, \qquad 0 \leq s \leq 1. \tag{3.43}$$

Using this technique, we again consider the inhomogeneous medium with hyperbolic tangent dependence on optical altitude, but this time we consider the parameters a and b known and the optical thickness T as the unknown. In trial one with $T = 0.9$, the estimate converges to the correct value of 1.0. Similarly, beginning with $T = 1.5$ or with $T = 0.5$, the correct value of T is obtained to within one part in 100,000.

3.5.3 Numerical results for albedo of a layered medium

In the trial for the estimation of all three parameters in the albedo function simultaneously from 49 intensity measurements, the successive approximations converge accurately to their correct values, as shown in Table 3.9.

Table 3.9. Successive Approximations of Three Parameters for a Layered Medium

Approximation	Lower Albedo $a - b$	Upper Albedo $a + b$	Thickness c
0	0.510000	0.69000	0.490000
1	0.420000	0.60520	0.503800
2	0.399929	0.59995	0.499602
3	0.399938	0.59994	0.499878

By now, we have seen how well quasilinearization works that we take for granted the excellent results in Table 3.9. Let us analyze these estimates. First, as one may have noticed when the computed reflected intensities for this layered slab were presented in Fig. 2.14, these intensities were close to those for a homogeneous slab with albedo 0.5 (the average value). Since the albedo is 0.6 in the upper half of the slab, a homogeneous slab with albedo 0.55 may be thought to be a more representative equivalent for this layered slab.

On the other hand, these estimates of a, b and c (and from these the albedoes of the lower and upper layers and the thickness) in Table 3.9 indicate that there are sufficient differences between the two cases–i.e., layered medium with 0.4 and 0.6 albedoes and the homogeneous medium with albedo 0.55–that a better fit is the layered case found in Table 3.9. Our interpretation is that these intensities contain information sufficient to discern a layered medium from a homogeneous one.

3.5.4 Numerical results for a parabolic albedo profile

We consider an inhomogeneous medium having the parabolic profile for the albedo for single scattering,

$$\lambda(t) = 0.5 + at + bt^2, \qquad 0 \le t \le c \qquad (3.44)$$

where a and b are constants and the optical thickness is c. In the medium of interest, the values of the constants are

$$a = 2.0, \quad b = -2.0, \quad c = 1.0. \qquad (3.45)$$

We apply the quasilinearization method beginning with the initial estimates

$$a = 2.2, \quad b = -1.8, \quad c = 1.5. \qquad (3.46)$$

After only four iterations, the following improved estimates are obtained:

$$a = 1.99895, \quad b = -1.99824, \quad c = 1.004. \qquad (3.47)$$

In another trial with

$$a = 2.2, \quad b = -1.8, \quad c = 0.5, \qquad (3.48)$$

the solution diverges. Evidently these estimates were not "close enough."

In other computational experiments, noise is introduced into the measurements. We consider noise that is alternatively 5% higher or lower in the measurements. Even with 5% noise, the estimates are quite good:

$$a = 1.91, \quad b = -1.85, \quad c = 1.02. \qquad (3.49)$$

However, if instead of 49 measurements, only 7 measurements are used (for a single input angle), the estimates range from 1.5 to 1.92 for a, -1.2 to -1.79 for b, and 1.07 to 1.3 for c.

3.5.5 Effect of optimizing criterion

In the above experiments, the least squares condition which is suitable for a Gaussian distribution of random noise has been used. To study the effect of the criterion on the estimates, we consider a minimax condition. The resultant problem changes from the solution of linear algebraic equations to the solution of a linear programming problem. The results are shown in Table 3.10.

The maximum deviations are respectively 0.02000 (as expected) and 0.02937. Obviously, the condition utilized must match the type of experimental errors expected.

Table 3.10. Numerical Results for Layered Medium using Minimax Criterion
(a) 2% equal magnitude errors
(b) 2% Gaussian errors

Type of Error	Estimate of a	Estimate of b	Estimate of c
(a)	1.99948	−1.99960	1.00001
(b)	1.76279	−1.67484	1.03852

3.5.6 Monte Carlo and the effect of noise on estimates

Noise in observations clearly affects quality of estimates. We have seen this in the discussions of associate memories, the parabolic albedo, and the effect of minimizing criterion.

The Monte Carlo method refers to the technique of carrying out computational experiments which are simulations in which errors are introduced into data and their effect is studied [13]. This includes operating on unaltered data using the methods of truncation, rounding, adding or subtracting fixed percentages of the unaltered data, as described above. A further important technique is that of generating random numbers according to an assumed distribution. Many software libraries contain such random number generators. (See, for example, *Numerical Recipes in C* [10]). The uniform and the Gaussian normal distributions are the ones most frequently used, and algorithms have been developed to generate numbers which represent samples from these distributions. Thus, frequently the Monte Carlo method refers to the sequence of obtaining simulated noisy data by "rolling the dice" to generate a random number and adding it to an observation. These noisy data are processed and analyzed.

To generate a noisy observation, do the following:

1. Obtain an accurate observation, a
2. Generate a random number, r
3. Input value of percent error, p
4. Compute noisy observation, n, $n = a(1 + 0.01pr)$.

Let us consider the estimation of source distributions from noisy measurements of upwardly emergent radiation. See Section 2.5.1 for the treatment of the direct problem. Let the source distribution with altitude y in a slab of thickness have the form

$$g(y) = 0.5[\tanh 10(y - a) - \tanh 10(y - b)], \quad 0 \le y \le 1.5,$$

where a and b are constants to be estimated. The observations are the emergent intensities,

$$e_i(1.5) \sim b_i, \quad i = 1, 2, \ldots, 7,$$

where the observations are made at the angles whose cosines are the points of the Gaussian quadrature formula of order 7, i.e., 88.5,...,13.0 degrees.

We wish to use quasilinearization to estimate the constants a and b and therefore the source distribution. We impose the optimizing criterion to be the least squares one, i.e., minimize the sum of squares of deviations, S, where

$$S = \sum_{i=1}^{7} [e_i(1.5) - b_i]^2.$$

We use the least squares criterion because in this simulated experiment we say that the errors are randomly distributed according to the Gaussian distribution. The least squares method is generally attributed to Gauss.

When the equations for determining the optimal set of parameters using quasilinearization are worked out, they turn out (not surprisingly) to have the form of linear algebraic equations. This means that, apart from the forming of the matrix of coefficients and the right hand side, the method of solution remains the same as before.

In our simulated experiments, the true values of a and b are

$$a = 0.4 \quad \text{and} \quad b = 0.7.$$

In the first experiment with "true" observations, the estimates obtained are

$$a \simeq 0.398 \quad \text{and} \quad b \simeq 0.699.$$

These slight changes are due to limited word length in the computer and accumulation of roundoff errors in the computation.

In the next three experiments, errors of varying amounts are introduced into the observations. Seven random numbers with a Gaussian distribution with 0.0 mean (unbiased) and standard deviation 1.0 (unit deviation) are generated. These numbers are multiplied by $0.001=0.1\%$ times the correct intensities, in the first of the three noisy experiments. Taking a fixed percentage of the correct intensity makes it such that the relative noise is the same in each observation, namely 0.1%. The other scaling, i.e., by multiplying by a random number, means that samples are taken from the distribution of interest. We find that these errors lead to errors in estimates of -1.7% and -0.69% in a and b, respectively, and an error in the estimate of the source strength at altitude $t = 0.5$ of $g(0.5) = +1.4\%$. The sum S is 0.1×10^{-5}.

This experiment is repeated with 1% and 2% errors. The results are summarized in Table 3.11. Observe how the estimates degrade with increasing errors in measurements. This makes for a fascinating study in informational analysis. Computational experiments like these are used in the planning of actual experiments.

For those contemplating carrying out a Monte Carlo computational study of the effect of errors on estimates, they are advised to go slowly, in phase-wise fashion. In Phase I, produce the computational solution of the direct problem, as discussed in Chapter 2, making sure that the results are correct. Vary the

Table 3.11. Errors in Observations and Errors in Estimates

% error in observations	% error in a	% error in b	% error in $g(1.5)$	Sum S
0.0	−0.4	−0.14	+0.35	0.2×10^{-6}
0.1	−1.7	−0.69	+1.4	0.1×10^{-5}
1.0	−14.	−5.8	+6.4	0.8×10^{-4}
2.0	−29.5	−12.	+3.9	0.3×10^{-3}

number of quadrature points and the step size of integration. Produce simulated accurate measurements. In Phase II, estimate the unknown parameters using the accurate observations from Phase I. The estimates should be close to their true values. In Phase III, the computer program must be able to generate and handle multiple sets of noisy observations for the Monte Carlo runs and to collect statistical data to summarize the results. Keep free the number of Monte Carlo trials until the number of trials required for an error analysis is determined. This may be done by studying how the statistics change as the number of trials is increased. The larger the noise, generally, the greater the number of trials needed. Be prepared to run hundreds or thousands of trials. Design a set of experiments in which the noise is at first small, about 0.1%. If the results are satisfactory, then increase the noise gradually up to the desired maximum amount. Analyze the results and draw conclusions. The analysis of the effect of errors in measurements on errors in estimates can also be made without actually carrying out numerous computational experiments necessary in using Monte Carlo. In Reference [16], the authors show how to use the FEED procedure to compute all of the partial derivatives necessary to evaluate the statistics of the noisy effect. Also in the same paper, the topic of model noise, i.e., errors in modeling, is considered. The system parameter estimates can be sharpened up considerably when the appropriate model is used.

3.6 Other inversion techniques

The dynamic programming technique [3], which formulates and solves optimization problems for multistage decision processes, is also suitable for inverse problems in radiative transfer. The absorbing properties of an inhomogeneous medium were estimated on the basis of knowing the internal intensities of radiation at points in the medium [1].

As we are aware, the reflected or transmitted intensity of a slab illuminated by external radiation is expressible as an integral over optical altitude of a weighted source function. For a semi-infinite slab, this integral is the Laplace transform of the source function. Some inverse problems take the form of the inversion of Laplace transforms so as to estimate the corresponding unknown source functions. A numerical technique presented in [15] is also suitable for the estimation of internal distributions of emitting sources on the basis of radiation emerging from the top of the bottom.

3.7 Discussion

The computational results presented here are representative of the myriad opportunities for learning more about a situation of interest through inverse problems. Other inverse problems include temperature profile estimation through atmospheric sounding, estimation of phase functions for anisotropic scattering, and estimating properties of reflecting surfaces and source distributions. The reader is encouraged to study the other chapters in this book, and to refer to the original publications for the details.

References

1. H. Kagiwada, *System Identification: Methods and Applications*, Addison-Wesley Publishing Company, Reading, Mass., 1974.

2. A. P. Wang and S. Ueno, "Invariant Imbedding and Inverse Problems with Applications to Radiative Transfer," to appear in *Computer Math. Applic.*

3. R. Bellman *Dynamic Programming*, Princeton University Press, Princeton, 1959.

4. R. Kalaba, "On Nonlinear Differential Equations, the Maximum Operation, and Monotone Convergence," *J. Math. and Mech.*, 1959, pp. 519–574.

5. R. Bellman, H. Kagiwada, and R. Kalaba, "Orbit Determination as a Multi-Point Boundary-Value Problem," *Proc. National Academy of Sciences*, Vol. 48, 1962, pp. 1327–1329.

6. R. Bellman, H. Kagiwada, R. Kalaba, and S. Ueno, "Inverse Problems in Radiative Transfer: Layered Media," *Icarus*, Vol. 4, 1965, pp. 119–126.

7. T. Kohonen, *Self-Organization and Associative Memory*, Springer-Verlag, New York, 1987.

8. H. Kagiwada, R. Kalaba, S. Timko, and S. Ueno, "Associative Memories for System Identification: Inverse Problems in Remote Sensing," *Mathematical and Computational Modelling*, Vol. 14, 1990, pp. 200–202.

9. D. Marr and T. Poggio, "A Computational Theory of Human Stero Vision," *Proceedings of the Royal Society London*, Series B204, 1979, pp. 301–328.

10. W. Press, et al., *Numerical Recipes in C*, Cambridge University Press, Cambridge, 1992.

11. R. E. Wengert, "A Simple Automatic Derivative Evaluation Program," *Communications of the Assoc. for Computing Machinery*, Vol. 7, 1964, pp. 463–464.

12. R. Bellman, H. Kagiwada and R. Kalaba, "Wengert's Numerical Method for Partial Derivatives, Orbit Determination, and Quasilinearization," *Comm. Assoc. Comput. Mach.*, Vol. 8, 1965, pp. 231–232.

13. H. Kagiwada, R. Kalaba, N. Rasakhoo, and K. Spingarn, *Numerical Derivatives and Nonlinear Analysis*, Plenum Press, New York, 1986.

14. R. W. Hamming, *Numerical Methods for Scientists and Engineers*, Dover Publications, New York, 1973.

15. R. Bellman, H. Kagiwada and R. Kalaba, "Numerical Inversion of Laplace Transforms and Some Inverse Problems in Radiative Transfer," *J. Atmosph. Sci.*, Vol. 23, 1966, pp. 555–559.

16. H. H. Kagiwada, J. K. Kagiwada and R. E. Kalaba, "Precision Passive Ranging," *Comput. and Math. with Appl.*, Vol. 26, 1993, pp. 89–96.

4. Anisotropic Scattering

Terrestrial atmospheres are neither isotropic nor homogeneous. In this chapter, we first take up a model of anisotropic scattering in an inhomogeneous slab, assuming that the local scattering is described by a function of only the incident and scattered polar angles and the vertical coordinate. This function is readily parameterized to approximate phase functions that vary from isotropic to highly elongated and anisotropic. Next we discuss a phase function that can be expanded in a series of Legendre polynomials and Cauchy problems for reflection and transmission functions that are also expanded in a similar series. Such a series approximation is appropriate for mildly anisotropic phase functions. Then we consider some inverse problems of estimating phase functions based on radiation measurements. There are also some approximate formulas that are rather useful and accurate. Finally, we take up diffuse reflection in a three-dimensional medium. Numerical results are presented in graphs and tables.

4.1 Introduction

Suppose that a pencil of radiation is going in the direction whose polar angle is $\theta = \arccos v$, measured from the upward vertical, and the azimuth angle is ϕ, measured from a specific axis. When it is anisotropically scattered from the direction $\Omega = (\theta, \phi)$ into the direction $\Omega' = (\theta', \phi')$, the process is that of interaction with the medium followed by scattering into the direction Ω' with probability p described by the phase function, p,

$$p(\Omega, \Omega') = p(v, \phi, v', \phi'), \quad -1 \le v, v' \le +1, \quad 0 \le \phi, \phi' \le 2\pi, \quad (4.1)$$

which may depend upon time, spatial coordinates, and other physical parameters which are not expressed here. When scattering is isotropic, then p is equal to one. The phase function is normalized so that

$$(4\pi)^{-1} \int_{2\pi} p(\Omega, \Omega') 2\pi \sin \Theta d\Theta = (4\pi)^{-1} \int_{-1}^{+1} \int_0^{2\pi} p(v, \phi, v', \phi') dv' d\phi' = 1,$$
$$(4.2)$$

and Θ, the angle between the two directions, is given by

$$\cos \Theta = \cos \theta \cos \theta' + \sin \theta \sin \theta' \cos(\phi - \phi'). \tag{4.3}$$

The local reciprocity principle,

$$p(v, \phi, v', \phi') = p(v', \phi', v, \phi), \tag{4.4}$$

$$p(-v, \phi, v', \phi') = p(v, \phi, -v', \phi'), \tag{4.5}$$

indicates that the paths may be reversed and that there is invariance under an interchange of the upward and downward directions. Some phase functions are very peaked in the forward direction, others have structure in different directions.

4.1.1 One-dimensional reflection function

We begin the study of anisotropic scattering in inhomogeneous media by considering the "one-dimensional reflection function." In this model, the scattering properties of the atmosphere vary only vertically with z, the optical altitude. The albedo for single scattering is $\lambda(z)$, and the phase function is $p(z, \Omega, \Omega')$. When uniform monodirectional radiation is incident on the top surface with incident flux π, the intensity r of the diffusely reflected radiation from a slab of optical thickness z is

$$r(z, v, \phi, u, \phi_0) = S(z, v, \phi, u, \phi_0)/4v, \qquad 0 < u, v \le 1. \tag{4.6}$$

The invariant imbedding equations for this S function are obtained using the method of Chapter 2. Let us add a thin layer of thickness Δ to the top of the slab of thickness z. In a small volume at the top with unit horizontal area and height Δ, the effect of the thin layer is to (1) decrease the diffusely reflected radiation due to absorption of incident and emerging radiation; (2) allow incoming radiation to singly scatter into the emergent direction (v, ϕ); (3) radiation at the top that would be emerging in the direction $(v', \phi')dv'd\phi'$ interacts with the medium, is scattered into the direction (v, ϕ) and then emerges; (4) incoming radiation is singly scattered down into the medium from which it is diffusely reflected; (5) incoming radiation is diffusely reflected into upward directions (v'', ϕ''), singly scattered into downward directions (v', ϕ'), and diffusely reflected into upward direction (v, ϕ). In the limit as the thickness of the added layer goes to zero, the following integro-differential equation is obtained:

$$
\begin{aligned}
\partial S/\partial z \;=\; & -(v^{-1} + u^{-1})S + \lambda(z)\{p(z, v, \phi, -u, \phi_0) \tag{4.7}\\
& + (4\pi)^{-1} \int_0^1 \int_0^{2\pi} p(z, v, \phi, v', \phi')S(z, v', \phi', u, \phi_0)d\phi'dv'/v' \\
& + (4\pi)^{-1} \int_0^1 \int_0^{2\pi} S(z, v, \phi, v', \phi')p(z, -v', \phi', -u, \phi_0)d\phi'dv'/v'
\end{aligned}
$$

$$+ \ (16\pi^2)^{-1} \int_0^1 \int_0^{2\pi} \int_0^1 \int_0^{2\pi}$$

$$\cdot S(z, v, \phi, v', \phi')p(z, -v', \phi', v'', \phi'')d\phi' dv'/v'$$

$$\cdot S(z, v'', \phi'', u, \phi_0)d\phi'' dv''/v''\},$$

for $0 < z$. If the underlying surface is a Lambert reflector, the initial condition is

$$S(0, v, \phi, u, \phi_0) = 4A(x, y)uv. \tag{4.8}$$

For phase functions that are nearly isotropic, such as the Rayleigh phase function, the method of Gaussian quadrature for the approximation of integrals may be useful, and the reduction to a system of ordinary differential equations is similar to that of Chapter 2. Other treatments are indicated in the sections to follow.

4.1.2 One-dimensional transmission function

In a similar manner, we obtain the differential-integral equation for the diffusely transmitted intensity,

$$T(z, v, \phi, u, \phi_0)/4v, \qquad 0 < u, v \le 1, \tag{4.9}$$

where v is the cosine of the polar angle measured from the downward vertical. For the T function, the terms on the right are the effects of the thin layer to (1) decrease the diffusely transmitted radiation due to absorption of incident radiation; (2) allow incoming radiation to singly scatter into the emergent direction $(-v, \phi)$ and be attenuated through the medium before emerging from the bottom; (3) radiation at the top that would be emerging in the direction $(v', \phi')dv'd\phi'$ interacts with the medium, is scattered into the direction $(-v, \phi)$ and is attenuated before emerging from the bottom; (4) incoming radiation is scattered down into the medium from which it is diffusely transmitted; and (5) incoming radiation is diffusely reflected into upward directions (v', ϕ'), singly scattered into downward directions $(-v', \phi')$, and is diffusely transmitted to emerge from the bottom in the direction $(-v, \phi)$. The equations of the Cauchy problem are

$$\partial T/\partial z \ = \ -v^{-1}T + \lambda(z)\bigg\{ \exp(-z/v)p(z, -v, \phi, -u, \phi_0)$$

$$+ \ (4\pi)^{-1} \int_0^1 \int_0^{2\pi} \exp(-z/v)p(z, -v, \phi, v', \phi') \tag{4.10}$$

$$\cdot S(z, v', \phi', u, \phi_0)d\phi' dv'/v'$$

$$+ \ (4\pi)^{-1} \int_0^1 \int_0^{2\pi}$$

$$\cdot \exp(-z/v)T(z,v,\phi,v',\phi')p(z,-v',\phi',-u,\phi_0)d\phi'dv'/v'$$

$$+ \quad (16\pi^2)^{-1}\int_0^1\int_0^{2\pi}\int_0^1\int_0^{2\pi}$$

$$\cdot \exp(-z/v)T(z,v,\phi,v',\phi')p(z,-v',\phi',v'',\phi'')d\phi'dv'/v'$$

$$\cdot S(z,v'',\phi'',u,\phi_0)d\phi''dv''/v''\Big\},$$

for $0 < z$ and

$$T(0,v,\phi,u,\phi_0) = 0. \tag{4.11}$$

The one-dimensional S and T functions for anisotropic scattering have two additional variables compared with the one-dimensional S and T functions of isotropic scattering. Consequently, the solution of these equations is made more challenging. Still, for phase functions such as the Rayleigh phase function that are not so different from isotropic, Gaussian quadrature and other numerical techniques used previously can be applied. Other ways of treating anisotropic scattering are considered in the following sections.

The last four equations may be written more compactly for the functions

$$S(z,\Omega,\Omega_0) = S(z,v,\phi,u,\phi_0), \tag{4.12}$$

$$T(z,\Omega,\Omega_0) = T(z,v,\phi,u,\phi_0), \tag{4.13}$$

$$
\begin{aligned}
\partial S/\partial z &= -(v^{-1}+u^{-1})S + \lambda(z)\{p(z,\Omega,-\Omega_0) \\
&+ \quad (4\pi)^{-1}\int_{2\pi} p(z,\Omega,\Omega')S(z,\Omega',\Omega_0)d\Omega'/v' \\
&+ \quad (4\pi)^{-1}\int_{2\pi} S(z,\Omega,\Omega')p(z,-\Omega',-\Omega_0)d\Omega'/v' \\
&+ \quad (16\pi^2)^{-1}\int_{2\pi}\int_{2\pi} S(z,\Omega,\Omega')p(z,-\Omega',\Omega'')d\Omega'/v' \\
&\quad \cdot S(z,\Omega'',\Omega_0)d\Omega''/v''\},
\end{aligned}
\tag{4.14}
$$

$$S(0,\Omega,\Omega_0) = 4A(x,y)uv, \tag{4.15}$$

$$
\begin{aligned}
\partial T/\partial z \;=\; & -v^{-1}T + \lambda(z)\{\exp(-z/v)p(z,-\Omega,-\Omega_0) \\
+\; & (4\pi)^{-1}\int_{2\pi} \exp(-z/v)p(z,-\Omega,\Omega')S(z,\Omega',\Omega_0)d\Omega'/v' \quad (4.16) \\
+\; & (4\pi)^{-1}\int_{2\pi} \exp(-z/v)T(z,\Omega,\Omega')p(z,-\Omega',-\Omega_0)d\Omega'/v' \\
+\; & (16\pi^2)^{-1}\int_{2\pi}\int_{2\pi} \exp(-z/v)T(z,\Omega,\Omega')p(z,-\Omega',-\Omega'')d\Omega'/v' \\
& \cdot S(z,\Omega'',\Omega_0)d\Omega''/v''\},
\end{aligned}
$$

$$
T(0,\Omega,\Omega_0) = 0. \tag{4.17}
$$

4.2 Phase function dependent on polar angles

4.2.1 Basic equations

Suppose that the phase function under consideration has a form,

$$
p(\cos\theta) = k(b - \cos\theta)^{-1}, \qquad b > 1, \tag{4.18}
$$

where θ is the angle between the incident and scattered directions. This function can be written as

$$
c(v,u) = \lambda k2[(b - uv)^2 - (1 - u^2)(1 - v^2)]^{-1/2}, \tag{4.19}
$$

where k satisfies an appropriate normalization condition. We can approximate a stronger and stronger forward scattering by choosing a value of b closer and closer to unity, which is useful in many studies. We shall consider models in which the phase function is expressed as $c(v,u)$,

$$
\begin{aligned}
c(v,u) = \;& \text{the fraction of radiation that is scattered from} \\
& \text{the direction whose cosine is } u \text{ to the} \tag{4.20} \\
& \text{direction whose cosine is } v.
\end{aligned}
$$

The Cauchy problem for the scattering function is

$$
\begin{aligned}
S_x(v,u,x) \;=\; & -(u^{-1} + v^{-1})S(v,u,x) + 2c(v,-u) \\
& + \int_0^1 c(v,v')S(v',u,x)dv'/v' \\
& + 2\int_0^1 S(v,v',x)(dv'/v')[c(-v',-u)/2 \tag{4.21} \\
& + \int_0^1 (dv''/v'')c(-v',v'')S(v'',u,x)/4], \\
& x \geq 0, \quad 0 \leq v, \quad u \leq 1,
\end{aligned}
$$

$$S(v, u, 0) = 0, \qquad (4.22)$$

when the surface below is a perfect absorber.

Observe that the scattering function in this anisotropic model has the same arguments as that for isotropic scattering. This means that we can study effects of strong anisotropic scattering with computational ease.

4.2.2 Computation

The computational method is the same as before, namely, discretizing the scattering function by evaluating it at the N points of a Gaussian quadrature formula, then solving the Cauchy problem by numerical integration.

For the example given above, the constant k is

$$k = 2[\log(b + 1)/(b - 1)]^{-1}. \qquad (4.23)$$

In the following we set $\lambda = 1$. We choose $b = 1.1$, for a front-to-back scattering ratio of 20 to 1, and $N = 7$ with an Adams-Moulton fourth-order integration method. Five accurate digits are obtained in the computation of the scattering function. Next, choosing $b = 1.01$, a value very close to unity, we obtain three accurate figures. When we set $b = 1.001$, extremely close to unity, we must use $N = 15$ in our quadrature formula and a small integration step size of 0.001. Even so, we only obtain one accurate figure. This last case has a huge front-to-back scattering ratio of about 2,000, so these low approximations are quite expected. This approach seems to be quite useful for studying the effects of anisotropic scattering in homogeneous or, by extension to, inhomogeneous slabs.

4.3 Phase function expandable in Legendre polynomials

4.3.1 Basic equations

Let $p(\cos \alpha)$ be the phase function, where α is the scattering angle between an incident ray and a ray resulting from an elementary act of scattering. Let λ be the albedo for single scattering $(0 < \lambda \le 1)$. The quantity $\lambda p(\cos \alpha) d\Omega/4\pi$ is the fraction of absorbed energy which is scattered into a solid angle of magnitude $d\Omega$ around α. The phase function is normalized so that

$$\frac{1}{4\pi} \int_0^\pi p(\cos \alpha) 2\pi \sin \alpha \, d\alpha = 1. \qquad (4.24)$$

Let monodirectional flux of radiation be uniformly incident on the top of a horizontal, plane-parallel medium. The incident flux is π per unit area normal to the direction of incidence, and its polar angle is arc cosine $(-u)$ measured from the upward vertical $(0 < u \le 1)$. Let all polar angles be measured from the upward vertical, and let the azimuth angle be measured such that it is zero for the incident flux. The optical thickness of the slab is x.

The intensity of the diffusely reflected (multiply scattered) flux emerging from the top of the slab in the direction whose polar angle is arc cosine v ($0 < v \leq 1$) and whose azimuth is φ ($0 \leq \varphi \leq 2\pi$) is defined to be $r(x; v, \varphi; u, \varphi_0)$, where φ_0 is written for the incident azimuth. The scattering function S is introduced,

$$r(x; v, \varphi; u, \varphi_0) = \frac{S(x; v, \varphi; u, \varphi_0)}{4v}.$$

By the methods of the previous chapters and sections, one can obtain the differential-integral equation for S,

$$
\frac{\partial S(x; v, \varphi; u, \varphi_0)}{\partial x} + \left(\frac{1}{v} + \frac{1}{u} \right) S = \lambda \Big\{ p(v, \varphi; -u, \varphi_0)
$$
$$
+ \frac{1}{4\pi} \int_0^1 \int_0^{2\pi} S(x; v', \varphi'; u, \varphi_0) p(u, \varphi; u', \varphi') \frac{dv'}{v'} d\varphi'
$$
$$
+ \frac{1}{4\pi} \int_0^1 \int_0^{2\pi} S(x; v, \varphi, v'', \varphi'') p(-v'', \varphi''; -u, \varphi_0) \frac{dv''}{v''} d\varphi''
$$
$$
+ \frac{1}{(4\pi)^2} \int_0^1 \int_0^{2\pi} \int_0^1 \int_0^{2\pi} S(x; v', \varphi'; u, \varphi_0) p(-v'', \varphi''; v', \varphi')
$$
$$
\cdot S(x; v, \varphi; v'', \varphi'') \frac{dv'}{v'} d\varphi' \frac{dv''}{v''} d\varphi'' \Big\}. \tag{4.25}
$$

In the limit as the thickness approaches zero, we have the condition

$$S(0; v, \varphi; u, \varphi_0) = 0, \tag{4.26}$$

when the lower boundary is a perfect absorber. This condition depends upon the assumption of the characteristics of the lower boundary. For example, in the case of a reflecting Lambert surface the right-hand side of Eq. (4.26) will be nonzero.

4.3.2 Expansion approximation

It is supposed that the phase function may be expanded in a series of Legendre polynomials [2] consisting of $M + 1$ terms, where M is finite and, for practical computational purposes, should be about 20 or less,

$$p(\cos \alpha) = \sum_{m=0}^{M} c_m P_m (\cos \alpha). \tag{4.27}$$

The addition theorem for Legendre functions is utilized in Eq. (4.27), and the S function is expanded in the form

$$S(x; v, \varphi; u, \varphi_0) = \sum_{m=0}^{M} S^{(m)}(x; v, u) \cos m(\varphi - \varphi_0). \qquad (4.28)$$

Each of the $S^{(m)}$ components then satisfies the equation

$$\frac{\partial S^{(m)}}{\partial x} + \left(\frac{1}{v} + \frac{1}{u} \right) S^{(m)} \qquad (4.29)$$

$$= \lambda(2 - \delta_{om}) \sum_{i=m}^{M} (-1)^{i+m} c_i \frac{(i - m)!}{(i + m)!} \psi_i^m(v) \psi_i^m(u)$$

for $m = 0, 1, \ldots, M$, where

$$\psi_i^m(\mu) = P_i^m(\mu) + \frac{(-1)^{i+m}}{2(2 - \delta_{om})} \int_0^1 S^{(m)}(x; v, v') P_i^m(v') \frac{d\mu'}{\mu'} \qquad (4.30)$$

for $m = 0, 1, 2, \ldots, M$ and $i = m, m+1, \ldots, M$; δ_{om} is the Kronecker delta function ($\delta_{om} = 1$ if $m = 0$; otherwise $\delta_{om} = 0$); and $P_i^m(\mu)$ is an associated Legendre polynomial. The initial conditions are

$$S^{(m)}(0; v, u) = 0, \qquad m = 0, 1, \ldots, M. \qquad (4.31)$$

This reduces the problem to determining a set of functions of three variables (thickness and two angles).

4.3.3 Computation

In order to produce a computational solution, we introduce the functions of one variable,

$$S_{ij}^{(m)}(x) = S^{(m)}(x; z_i, z_j), \qquad (4.32)$$

where $z_1, z_2, \ldots, z_N$ are the roots in ascending order of the shifted Legendre polynomial, $P_N^*(z) = P_N(1 - 2z)$. We replace the definite integral in Eq. (4.30) by a sum according to the Gaussian quadrature method, with weights $w_1, w_2, \ldots, w_N$. Then we have the system of ordinary differential equations for $S_{ij}^{(m)}(x)$,

$$\frac{dS_{ij}^{(m)}(x)}{dx} = -\left(\frac{1}{z_i} + \frac{1}{z_j} \right) S_{ij}^{(m)} + \lambda(2 - \delta_{om})$$

$$\cdot \sum_{k=m}^{M} (-1)^{k+m} \frac{(k - m)!}{(k + m)!} c_k \psi_{ki}^m \psi_{kj}^m \qquad (4.33)$$

$$(m = 0, 1, \ldots, M; i = 1, 2, \ldots, N; j = 1, 2, \ldots, N),$$

where

$$\psi_{kl}^m = P_k^m(z_l) + \frac{(-1)^{k+m}}{2(2-\delta_{om})} \sum_{j=1}^{N} S_{lj}^{(m)} P_k^m(z_j) \frac{w_j}{z_j}. \tag{4.34}$$

The initial conditions are

$$S_{ij}^{(m)}(0) = 0. \tag{4.35}$$

It should be borne in mind that the independent variable is the thickness x. Equations (4.33) and (4.34) are integrated numerically from $x = 0$, using the initial condition in Eq. (4.35) to $x = x_0$, the desired thickness. Solutions may be obtained along the way for any thickness between 0 and x_0.

There are $(M+1) \cdot N^2$ equations implied in Eq. (4.33). However, due to the symmetry property, $S_{ij}^{(m)} = S_{ji}^{(m)}$, only $(M+1) \cdot N \cdot (N+1)/2$ equations need to be integrated.

The reflected flux is therefore calculated as the solution of an initial-value problem for ordinary differential equations. This problem has been found to be stable and well-suited for implementation on high-speed computers. Accurate solutions are readily obtained on digital computers. However, the value of M should not be larger than about 20 in order that the solution of the system of ordinary differential equations be reasonably accurate. This means that highly peaked phase functions for strong anisotropy cannot be treated in this manner without additional care being taken to ensure accuracy in the results.

4.4 Estimation of Phase Function

4.4.1 Estimation problem

The problem of estimating the local phase function of a medium by using gross measurements of reflected radiation will now be treated. Although the method is general for any phase function having the form shown in Eq. (4.27), for the sake of definiteness we consider the case in which the phase function is

$$p(\cos \alpha) = 1 + a \cos \alpha. \tag{4.36}$$

In this case the coefficients in the phase expansion are

$$c_0 = 1, \quad c_1 = a, \quad M = 1. \tag{4.37}$$

It is the coefficient a that we wish to estimate. Let its true value be 1.0, the local scattering being greatest in the forward direction. Let the slab thickness be $x_0 = 0.2$, and albedo $\lambda = 1.0$. We use $N = 7$. Let measurements be given of the reflected intensity having the above input and output polar direction cosines, and $K = 7$ output azimuths, $\varphi_1 = 0°$, $\varphi_2 = 30°, \ldots, \varphi_7 = 180°$,

$$r(x_0; z_i, \varphi_k; z_j, 0°) \approx b_{ijk}$$

$$(i = 1, 2, \ldots, N; j = 1, 2, \ldots, N; k = 1, 2, \ldots, K).$$

The objective is to minimize the sum of the squares of the deviations between r, the solution calculated using the above equations, and the given observations

$$Q = \sum_{i,j,k} \{r(x_0; z_i, \varphi_k; z_j, 0°) - b_{ijk}\}^2 \tag{4.38}$$

through the proper choice of the parameter a.

Thus, the estimation problem is formulated as a nonlinear boundary-value problem. Previous experience shows us how to obtain the computational solution using quasilinearization, a successive approximation scheme for differential equations.

4.4.2 Computational results

First, values of reflected intensity are computed by the method described in Section 4.3, and the method is checked out in a number of cases. We consider the case $x_0 = 0.2$, $a = 1$, $\lambda = 1$, $N = 7$, $M = 1$, and for seven azimuth angles, $0°, 30°, \ldots, 180°$. A FORTRAN IV program is used, together with a fourth-order predictor-corrector integration method and an integration step length of $\Delta x_1 = 0.01$. Twenty steps are taken on the interval from $x = 0$ to $x = 0.2$ for the system of 58 differential equations, Eq. (4.33). Reflected intensities in 49 outgoing directions are computed for each of 7 incident directions, a total of 343 numbers. Figure 4.1 shows a plot of 49 of these intensities. The incident direction cosine is $-u = -0.5$, and there are seven polar angles of the outgoing intensities. The horizontal axis is the azimuth angle from the forward to the backward direction. Thus, we see that the greatest intensity of radiation occurs near grazing angles in the forward direction.

The large dots in Figure 4.1 are intensity values containing 10 percent noise. To each correct value q times the correct value has been added, where q is a Gaussian random number from a sequence having a mean of 0 and a standard deviation of $0.1 = 10$ percent. Such incorrect values are used in the experiments to be described below to simulate errors in measurement. Also given in Figure 4.1 is a table of the reflected intensities for the case of conservative isotropic scattering. These are approximately the same as the average values of r on the curves for the forward-scattering phase function.

Several computational experiments are performed. In three controlled experiments without any errors in the observations, but with wrong initial estimates ($a = 0.8$, 0.5 and 0.0), the refined estimate $a = 0.99999$ is obtained. Note that setting $a = 0$ implies isotropic scattering as the initial estimate.

In the second series of experiments, errors of 0.5, 1, 2, 5, and 10 percent are introduced into the observations of r, and estimates of the parameter a are obtained which are all less than approximately 1 percent in discrepancy. In the experiment with 10 percent error and with isotropic scattering as the initial estimate, the sequence of values of the coefficient a is as follows:

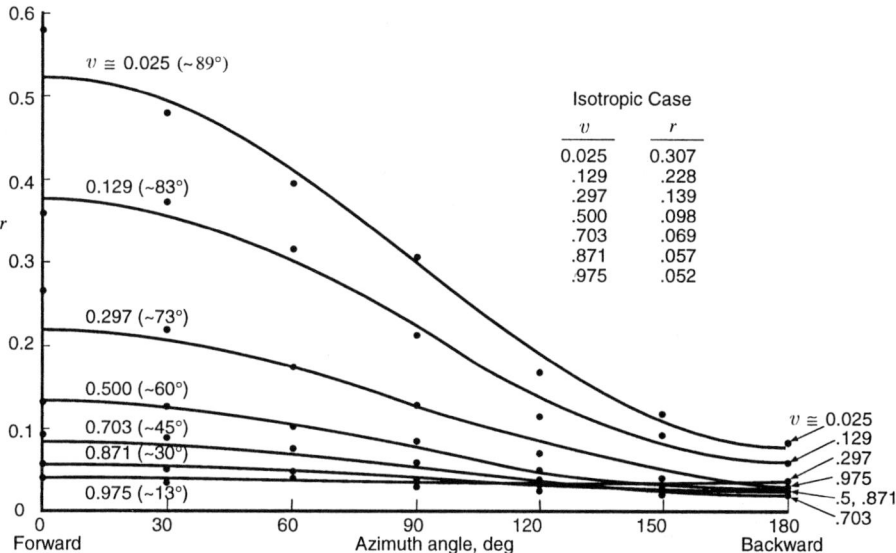

Figure 4.1. Reflection coefficient r, for phase function $p = 1 + \cos\alpha$, albedo 1, thickness 0.2, $-u = -0.5$.

$$a^0 = 0.0,$$
$$a^1 = 1.087,$$
$$a^2 = 1.014,$$
$$a^3 = 1.014.$$

Thus, even with large errors in the measurements and with a completely wrong initial estimate of the coefficient, the estimate of the coefficient is refined in only two steps to a discrepancy of only 1.4 percent.

4.5 Flux Equivalences

4.5.1 Introduction

As we have seen, the functional equations for radiative transfer can be quite formidable. Approximate formulas that reduce the computational effort are often welcome. After evaluating them against precision numerical results, they can be used to obtain quick values for certain quantities. In this section, we carry out a numerical comparison of some simple formulas against more accurate values. In a sense, these formulas represent simpler models of complex physical processes which, at least for the cases examined, do a good to excellent job of describing the situation.

4.5.2 Reflected and transmitted fluxes with isotropic scattering

Consider a horizontal, homogeneous, absorbing and isotropically scattering plane-parallel medium of finite optical thickness x. Monodirectional radiation with direction cosine u relative to the downward directed vertical is uniformly incident on the top surface of the slab. The case of normal incidence ($u = 1$) is of particular interest. The net flux per unit area normal to the incident rays is π. The albedo for single scattering is λ ($0 < \lambda \leq 1$). We shall here consider only the case of conservative scattering, $\lambda = 1$. The lower surface is a completely absorbing barrier.

Let $r(v, u, x)$ be the intensity of the multiply-scattered radiation which emerges from the top of the medium with direction cosine v relative to the upward directed vertical. Let $t(v, u, x)$ be the intensity of the diffusely transmitted radiation emerging from the bottom with direction cosine v relative to the downward vertical. The function $t(v, u, x)$ refers to radiation which has interacted one or more times with the medium. The intensity of the reduced incident radiation which is directly transmitted is $\frac{1}{2}\delta(u-v)\exp(-x/u)$, where $\delta(\ )$ denotes the Dirac delta function. This corresponds, of course, to a net reduced incident flux of $\pi u \exp(-x/u)$.

Reflected and transmitted fluxes computed via the invariant imbedding method are denoted $\rho_I(u, x)$ and $\tau_I(u, x)$ in the table that follows.

For isotropic scattering, the reflected flux is defined to be

$$\rho(u, x) \equiv 2\pi \int_0^1 r(z, u, x) z \, dz, \qquad (4.39)$$

the diffusely transmitted flux,

$$\tau(u, x) \equiv 2\pi \int_0^1 t(z, u, x) z \, dz, \qquad (4.40)$$

and the total transmitted flux is

$$\tau(u, x) + \pi u e^{-x/u}. \qquad (4.41)$$

The conservation law requires that

$$\pi u = \rho(u, x) + \tau(u, x) + \pi u e^{-x/u}. \qquad (4.42)$$

This serves as a convenient check on numerical values of ρ and τ.

4.5.3 Reflected and transmitted fluxes with Rayleigh scattering in a slab

Reflected and transmitted fluxes for Rayleigh slabs have been determined by Kahle, using a theory for singular integral equations. We refer the reader to her paper [9] and the references cited therein. The fluxes determined by Kahle are denoted $\rho_K(u, x)$ and $\tau_K(u, x)$. These fluxes could also have been obtained via the invariant imbedding equations discussed above.

4.5.4 Approximate formulas

Gavallas and Kagan [8] have developed formulas for reflected and transmitted fluxes for the case of conservative scattering and normal incidence. They assumed that the intensities have the form

$$A + B \cos \theta,$$

where θ is the polar angle. Their final expressions for the fluxes, obtained with the use of the Bubnov-Galerkin method, are

$$\rho_G(1, x) = \pi \frac{-1 + 3x + e^{-x}}{4 + 3x},$$

$$\tau_G(1, x) = \pi \frac{5 - 5e^{-x} - 3xe^{-x}}{4 + 3x}. \qquad (4.43)$$

Next we consider a simple model of isotropic scattering in a rod. Let a unit of energy per unit time be incident on the right end of a rod of length x. An analysis by invariant embedding leads to the equation for the reflected flux, $s(x)$,

$$s(x) = \frac{x}{x+2}. \tag{4.44}$$

To make the proper comparisons with the other models, we set the incident flux equal to πu and the effective length equal to x/u. Then the reflected flux is

$$\rho_0(u, x) = \pi u \frac{xu}{(x/u) + 2}$$

or

$$\rho_0(u, x) = \pi u \frac{x}{x + 2u}. \tag{4.45}$$

From the conservation Eq. (4.42), the transmitted flux is seen to be

$$\tau_0(u, x) = \pi u \left[\frac{2u}{x + 2u} - e^{-x/u} \right]. \tag{4.46}$$

4.5.5 Computational results

Computations for the conservative case ($\lambda = 1$) in slab geometry are carried out with the invariant imbedding equations for the slab. The order of the quadrature formula, N, and the step size of integration, Δx, are varied. The pairs of parameter values ($N, \Delta x$) are (3, 0.05), (3, 0.01), (5, 0.01), (7, 0.01), and (7, 0.005). The effect of changing the step size from 0.01 to 0.005 when $N = 7$ is a change, at most, of one unit in the fifth significant figure of intensity. The same is true for $N = 3$ and the two step sizes tested. Both hold for thicknesses up to 50. The overall agreement among all of the trials is 3 to 4 decimal places. The invariant imbedding numerical results quoted below are obtained with $N = 7$ and $\Delta x = 0.01$.

Eight curves of diffusely transmitted flux are shown in Figure 4.2 for the incident angles indicated. Note that the dots are plots of the values obtained by Kahle for Rayleigh scattering and incident angles 0, 60, and 88 (not 88.5) degrees. Kahle has also computed fluxes at $x = 100$. From this figure and by graphical interpolation, it is clear that the results coincide over all thicknesses up to 100. A similar conclusion can be drawn from Figure 4.3 for reflected flux.

The formulas of Gavallas and Kagan are excellent approximations to the reflected and transmitted fluxes on the interval $0 \leq x \leq 50$. The greatest discrepancy from the invariant imbedding calculation is 0.0102 at $x = 2$. For the most part, the discrepancy is in the order of 0.001.

A brief tabular survey of the comparisons among the four methods of determining the reflected and diffusely transmitted fluxes is presented in Tables 4.1 and 4.2.

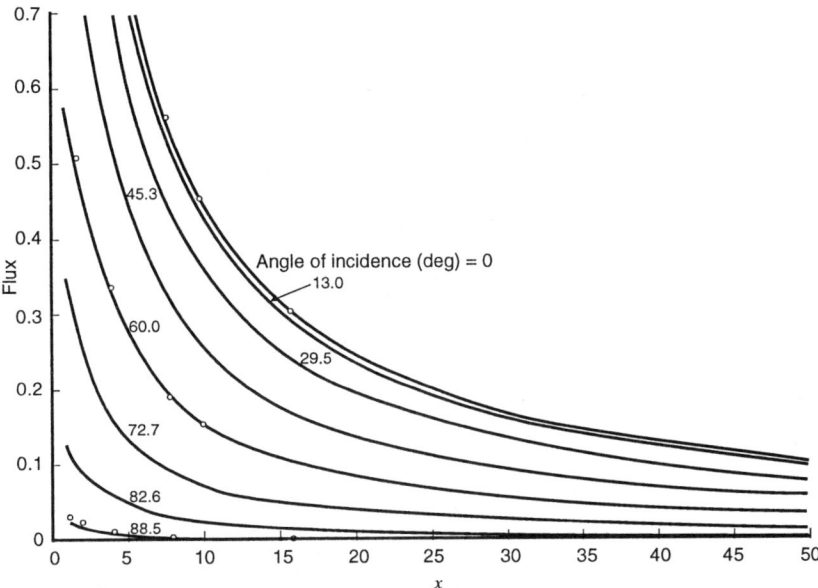

Figure 4.2. Fluxes diffusely transmitted through conservatively scattering slabs with various angles of incidence. The curves are for the isotropic scattering case, the dots for Rayleigh scattering.

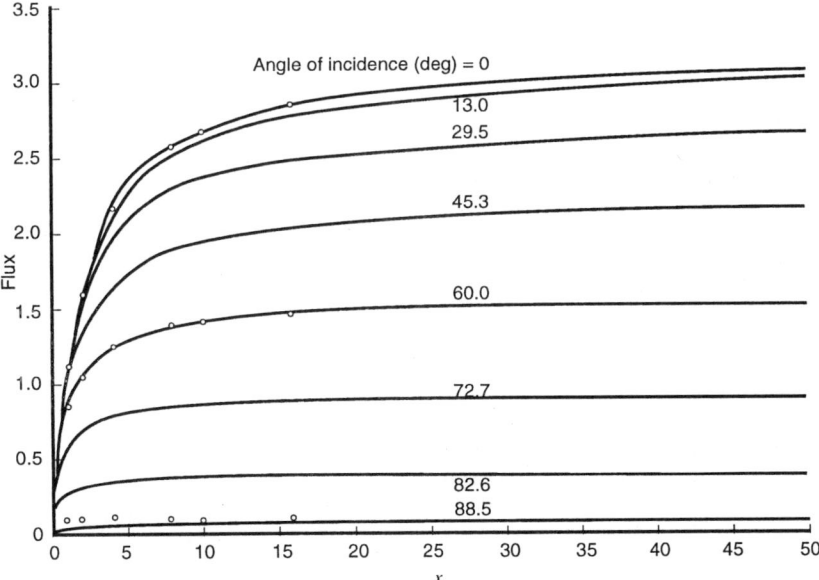

Figure 4.3. Fluxes reflected from conservatively scattering slabs with various angles of incidence. The curves are for the isotropic scattering case, the dots for Rayleigh scattering.

Table 4.1. A Comparison of Reflected Fluxes

x	$\cos^{-1} u(\deg)$	ρ_0	ρ_I	ρ_G	ρ_K
0.15	0	0.2193	0.2198	0.2194	0.2196
	60	0.2048	0.2054	—	0.2054
	88.5	0.0596	0.0446	—	0.0598[a]
1.0	0	1.0471	1.0723	1.0627	1.0687
	60	0.7854	0.7829	—	0.7842
	88.5	0.0760	0.0591	—	0.0813[a]
10.0	0	2.6179	2.6800	2.6796	2.6762
	60	1.4279	1.4110	—	1.4120
	88.5	0.0795	0.0756	—	0.1038[a]
50.0	0	3.0206	3.0400	3.0396	—
	60	1.5400	1.5357	—	—
	88.5	0.0798	0.0790	—	—

[a] $\cos^{-1} u = 88.0$ degrees.

Table 4.2. A Comparison of Diffusely Transmitted Fluxes

x	$\cos^{-1} u(\deg)$	τ_0	τ_I	τ_G	τ_K
0.15	0	0.2184	0.2178	0.2182	0.2179
	60	0.2022	0.2018	—	0.2017
	88.5	0.0200	0.0351	—	0.0484[a]
1.0	0	0.9387	0.9136	0.9232	0.9172
	60	0.5728	0.5754	—	0.5740
	88.5	0.0039	0.0208	—	0.0284[a]
10.0	0	0.5235	0.4618	0.4619	0.4653
	60	0.1428	0.1599	—	0.1588
	88.5	0.0004	0.0044	—	0.0059[a]
50.0	0	0.1208	0.1031	0.1020	—
	60	0.0308	0.0357	—	—
	88.5	0.0001	0.0010	—	—

[a] $\cos^{-1} u = 88.0$ degrees.

4.6 Three-Dimensional Reflection and Transmission

4.6.1 Three-dimensional medium

In many applications of radiative transfer in terrestrial atmospheres, the reflecting properties of the underlying earth, such as land versus ocean, greatly affect the radiation fields. The interactions must be considered if space-based images are to be accurately processed to remove atmospheric effects and determine the structure of the surface features. This aspect of remote sensing is discussed in Chapters 7 and 8. As is abundantly clear, this is an extremely complex problem from mathematical and computational aspects. At the time of this writing high precision solutions have not been computed, and approximations are the order of the day.

The three-dimensional model has a finite, vertically inhomogeneous and anisotropically scattering atmosphere bounded below by the horizontally inhomogeneous albedo of the ground. The geometrical coordinates of the problem are horizontal optical coordinates (x, y) and vertical coordinates (optical altitude t or optical thickness z). The radiation fields of interest are actually functions of other variables as well; in this treatment, we have six variables. The layer is illuminated uniformly at the top by parallel rays of radiation of net flux πF per unit area normal to the direction of propagation, Ω_0. The direction Ω_0 represents the polar angle $\theta_0 = \cos^{-1} u$ and azimuth angle ϕ_0. Let us define the "three-dimensional reflection function,"

$r(z, x, y, \Omega, \Omega_0) =$ the intensity of radiation emerging from the top of the layer of thickness z at the point whose coordinates are (x, y) and going in a direction described by the angle Ω, the incident radiation being given by Ω_0,

where Ω represents the polar angle $\theta = \cos^{-1} v$ and azimuth angle ϕ. The anisotropic phase function $p(t, \Omega, \Omega')$ depends on the optical altitude t and the incident and scattered directions.

The surface at the bottom of the atmosphere is assumed to have an albedo of reflection that varies with x and y, as well as with angle. For the case of Lambert's law of reflection, the downwelling radiation is reflected isotropically from the surface with an albedo $A(x, y)$.

4.6.2 Reflection function

The scattering function $S(z, x, y, \Omega, \Omega_0)$ is introduced through the relation,

$$r(z, x, y, \Omega, \Omega_0) = FS(z, x, y, \Omega, \Omega_0)/4v \qquad (4.47)$$

and is a function of three geometrical variables and four angles. Through the invariant imbedding of the function r (or S) as the optical thickness is varied, the following initial value problem is obtained for $S = S(z, x, y, \Omega, \Omega_0)$:

$$\partial S/\partial z + (\tan\theta\cos\phi + \tan\theta_0\cos\phi_0)\partial S/\partial x$$

$$+(\tan\theta\sin\phi + \tan\theta_0\sin\phi_0)\partial S/\partial y$$

$$= -\alpha(z)(v^{-1} + v_0)S + \lambda(z)\Big[p(z,\Omega,-\Omega_0) \qquad (4.48)$$

$$+(4\pi)^{-1}\int_{2\pi} p(z,\Omega,\Omega')S(z,x,y,\Omega',\Omega_0)\frac{d\Omega'}{v'}$$

$$+(4\pi)^{-1}\int_{2\pi} S(z,x,y,\Omega,\Omega'')p(z,-\Omega'',-\Omega_0)\frac{d\Omega''}{v''}$$

$$+(16\pi^2)^{-1}\int_{-\infty}^{\infty}\int_{-\infty}^{\infty}\int_{2\pi}\int_{2\pi}$$

$$\cdot S(z,x-x',y-y',\Omega,\Omega')p(z,-\Omega',\Omega'')$$

$$\cdot S(z,x',y',\Omega'',\Omega_0)dx'dy'\frac{d\Omega'}{v'}\frac{d\Omega''}{v''}\Big]$$

for derivation of (4.43) see 9.2.

Observe the three partial derivatives (of geometrical coordinates) and sextuple integrals: two over incident angles, two over outgoing angles, and two over horizontal coordinates. This initial value problem simplifies for isotropic scattering and for homogeneous atmospheres, for completely absorbing surfaces, and for those with only x or y variations in reflecting properties. This last problem may be suitable for studying the effect of coastal zones on atmospheric radiation, where the interface between land and sea may be considered to be a straight or curved line.

4.7 Concluding Remarks

The three-dimensional scattering function is a function of six variables in addition to the optical thickness. There are sextuple integrals to be evaluated, two of them over infinite intervals; this is not a straightforward computational matter. The computational solution has only been obtained for problems of reduced dimensionality or in various approximations. Earlier in this chapter we were able to solve the problem under certain assumptions: a phase function dependent on polar angles only, an expansion of the solution with a few Legendre polynomials, and for anisotropic scattering phase functions that are nearly isotropic, such as Rayleigh scattering.

The three-dimensional anisotropic diffuse reflection and transmission functions are important in terrestrial remote sensing, as discussed in the next chapter. Although we do not take it up in this book, another remote sensing application is the probing of patients' arteries using laser beams to estimate the state of their cardiovascular health. For such problems of terrestrial and medical remote sensing, the consideration of the nature of the incident radiation as a narrow beam may be necessary. Please refer to Chapter 10 on the "searchlight problem" with anisotropic scattering. For more details on the topics in this chapter, see [1]–[11].

References

1.   V. V. Sobolev, *Dokl. Akad. Nauk SSSR*, Vol. 179, 1968, p. 41 [*Sov. Phys.-Dokl.*, Vol. 13, 1968, p. 180].

2.   S. Chandrasekhar, *Radiative Transfer*, Dover, New York, 1960.

3.   A. Leonard and T. Mullikin, *J. Math. Phys.*, Vol. 5, 1964, p. 399.

4.   H. Kagiwada and R. Kalaba, "Estimation of Local Anisotropic Scattering Properties Using Measurements of Multiply-Scattered Radiation," *J. Quant. Spect. Rad. Transfer*, Vol. 7, 1967, pp. 295–303.

5.   H. Kagiwada and R. Kalaba, "Exact Solution of a Family of Integral Equations of Anisotropic Scattering," *J. Math. Phys.*, Vol. 11, 1970, pp. 1575–1578.

6.   R. Bellman, H. Kagiwada, R. Kalaba and S. Ueno, "Some Mathematical Aspects of Multiple Scattering in a Finite Inhomogeneous Slab with Anisotropic Scattering," *Publ. Astron. Soc. Japan*, Vol. 22, 1970, pp. 75–83.

7.   H. Kagiwada, R. Kalaba and R. Segerblom, "Flux Equivalences of Isotropic, Rayleigh, and Other Scattering Models," *J. Comp. Phys.*, Vol. 3, 1968, pp. 159–166.

8.   L. A. Gavallas and Yu. M. Kagan, "Multiple Scattering of Slow Electrons in a Gas," Part I. *Izv. Vuz Fiz.*, Vol. 2, 1965, pp. 52–56 [English transl.: *Soviet Phys. J.*, Vol. 2, 1965, pp. 33–36].

9.   A. Kahle, "Global Radiation Emerging from a Rayleigh Scattering Atmosphere of Large Optical Thickness," *Astrophys. J.*, Vol. 151, 1968, pp. 637–645.

10.   R. Bellman, R. Kalaba and S. Ueno, "Invariant Imbedding and Diffuse Reflection from a Two Dimensional Flat Layer," *Icarus*, Vol. 1, 1963, pp. 297–303.

11.   A. P. Wang, "Multi-scattering of Anisotropic Radiative Transfer Through Layer of Cloud," *J. of Applied Sci. Research*, Vol. 23, 1970, pp. 221–236.

5. Finite Orders of Scattering

When remote sensing of the atmosphere is performed at wavelengths whose optical thickness is small or when the albedo for single scattering is small, then reflected, transmitted, and internal intensities can be computed through an approximate method which enumerates the intensity fields up to a finite number of scatterings. Results are also useful for approximating the general problem and for understanding the physical effects of the order of scattering.

5.1 Introduction

In the theory of radiative transfer, when a parallel beam of radiation of constant net flux πF impinges uniformly on a plane-parallel, homogeneous, isotropically scattering atmosphere of optical thickness $x > 0$, the following functions play an important role: $I(t, \pm v)$ represents the specific intensity of radiation existing at level t $(0 \leq t \leq x)$, traveling at an angle arc $\cos(v)$ $(0 < v \leq 1)$ with the downward normal at that point; $S(x; v, u)$ is called the scattering function by which the reflected intensity at the top $t = x$ is expressed in the form

$$I(x, -v) = (F/4v)S(x; v, u),$$

where u and v represent, respectively, the cosine of the angles of incidence and reflection; and (3) $T(x; v, u)$ is called the transmission function by which the diffusely transmitted specific intensity at the bottom $t = 0$ is expressed in the form

$$I(0, +v) = (F/4v)T(x; v, u).$$

The knowledge of these functions is vital for determining the amount of energy delivered to the medium at different depths after one or more scattering processes, not only in radiative transfer [3, 5, 6, 7, 8, 9] but also in radiation dosimetry [1, 2, 4]. Hence, in what follows, we shall concern ourselves with these functions, specifying the number n of scattering processes during the passage of the radiation through the slab.

Consider a monodirectional beam of radiation of flux πF per unit area normal to itself, which is uniformly incident in a direction u $(0 < u \leq 1)$ on the top $t = x$ of a homogeneous target slab of optical thickness x with isotropic and coherent scattering properties, where u is a cosine of the angle of incidence measured from the downward normal. Let the specific intensity of radiation which has undergone n scatterings $(n \geq 1)$ present in a direction v $(0 < v \leq 1)$ with respect to the downward normal at level t $(0 \leq t \leq x)$ be denoted by $I(n, t, v)$.

The equation of transfer in the diffuse radiation field takes the form

$$v\frac{dI}{dt}(n, t, v) = I(n, t, v) - \frac{\lambda}{2} \int_{-1}^{+1} I(n - 1, t, w) dw \qquad (5.1)$$
$$-\frac{\lambda F}{4} e^{-(x-t)/u} \delta(n, 1),$$

where $n \geq 1, \lambda$ is an albedo for single scattering, representing the ratio of the intensity after and before each elementary act of scattering, and $\delta(n, 1)$ is the Kronecker delta function, which is equal to unity for $n = 1$ and zero otherwise. The case of $n = 0$ does not appear in the diffuse radiation field. Equation (5.1) should be solved subject to the boundary conditions

$$I(n, x, +v) = 0 \quad \text{and} \quad I(n, 0, -v) = 0. \qquad (5.2)$$

The last term of the exponential form on the right-hand side of Eq. (5.1) represents the contribution due to a narrow pencil of radiation which has penetrated to the level t without having undergone any scattering processes. In other words, this term acts as if it were an emitting radiation source with no incident radiation from the outside.

Allowing for the first several orders of successive scatterings, the principle of invariant imbedding appropriate to this case leads to the following expressions relating I- and S-functions:

$$I(n, t, -v) = (F/4v)e^{-(x-t)/u} S(n, t; v, u) \qquad (5.3)$$
$$+(1/2v) \sum_{i=1}^{n-1} \int_0^1 S(n - i, t; v, w) I(i, t, +w) \, dw,$$

where $S(n, x, v, u)$ $(n \geq 1, 0 \leq t \leq x, 0 < v, u \leq 1)$ is the scattering function of order n of radiation and $I(i, t, +w)$ is the downward intensity at level t for the ith order of scattering. Similarly, we have

$$I(n, t, +v) = (F/4v)T(n, x - t; v, u) \qquad (5.4)$$
$$+(1/2v) \sum_{i=1}^{n-1} \int_0^1 S(n - i, x - t; v, w) I(i, t, -w) dw,$$

where $T(n, x - t, v, u)$ $(n \geq 1, 0 \leq t \leq x, 0 < v, u \leq 1)$ is the transmission function of order n of radiation, and $I(i, t, -w)$ is the upward intensity of radiation at level t for the ith order of scattering.

5.2 Scattering and Transmission Functions of Finite Order

5.2.1 Finite order scattering functions

In what follows, with the aid of invariant imbedding, we derive the recurrence equations governing the finite order scattering and transmission functions. On differentiating eq. (5.3) with respect to t, passing to the limit $t = x$, and making use of eq. (5.1), after some minor arrangement of terms, we have (5.5).

$$\frac{\partial S}{\partial x}(n, x; v, u) + \left(\frac{1}{v} + \frac{1}{u}\right) S(n, x; v, u) \tag{5.5}$$

$$= \lambda \left[\delta(n, 1) + \frac{1}{2}\int_0^1 S(n-1, x; v, w)\frac{dw}{w} + \frac{1}{2}\int_0^1 S(n-1, x; w, u)\frac{dw}{w}\right.$$

$$\left. + \frac{1}{4}\sum_{m=1}^{n-1}\int_0^1 S(n-m, x; v, w)\frac{dw}{w}\int_0^1 S(m-1, x; w', u)\frac{dw'}{w'}\right],$$

where the diffusely reflected intensity $I(n, x, -v)$ is

$$I(n, x, -v) = (FS/4v)(n, x; v, u), \tag{5.6}$$

together with the initial condition

$$S(n, 0; v, u) = 0 \quad \text{and} \quad S(0, x; v, u) = 0. \tag{5.7}$$

This is a Cauchy problem for a system of equations with $n = 1, 2, \ldots, \infty$, or up to a maximum number of scatterings. To be practical, n would be at most about 10, and λ would be about 0.5 or less.

The above eq. (5.5) can also be straightforwardly derived without making use of the transfer eq. (5.1), by relating the reflected intensity by a slab of thickness x to that by a slab of thickness $x + \Delta$ where Δ is an infinitesimal length of thickness. This procedure can be done for each order of scattering function $S(n, x; v, u)$ by picturing the possible processes that can take place in increasing the thickness from x to $x + \Delta$. This is the standard procedure of the invariant imbedding method. Putting $n = 1$, in eq. (5.5), after some manipulations, we have

$$S(1, x; v, u) = \lambda vu\left[1 - \exp\left\{-x\left(\frac{1}{v} + \frac{1}{u}\right)\right\}\right]\bigg/(v + u), \quad x > 0. \tag{5.8}$$

Then, starting with eq. (5.8) for $n = 1$, and making use of eq. (5.5), we can successively compute the required scattering function of the nth order.

From the above equations, we can prove the reciprocity principle for the scattering function

$$S(n, x; v, u) = S(n, x; u, v). \tag{5.9}$$

On summing over all n, we get the invariant imbedding equation for the total scattering function,

$$\frac{\partial S}{\partial x}(x; v, u) + \left(\frac{1}{v} + \frac{1}{u}\right) S(x; v, u) \tag{5.10}$$

$$= \lambda \left[1 + \frac{1}{2} \int_0^1 S(x; v, w) \frac{dw}{w} + \frac{1}{2} \int_0^1 S(x; w, u) \frac{dw}{w} \right.$$

$$\left. + \frac{1}{4} \int_0^1 S(x; v, w) \frac{dw}{w} \int_0^1 S(x; w', u) \frac{dw'}{w'} \right],$$

where

$$S(x; v, u) = \sum_{n=1}^{\infty} S(n, x; v, u). \tag{5.11}$$

We introduce S^*,

$$S(n, x; v, u) = \lambda^n S^*(n, x; v, u), \tag{5.12}$$

where $S^*(n, x; v, u)$ is more convenient for numerical computation because it does not depend on the albedo for single scattering. By substituting the relation (5.12) into (5.5), we obtain the Cauchy problem for $S^*(n, x, v, u)$ that is identical to that for $S(n, x, v, u)$ (namely, (5.5) and (5.7)) except that there is no λ in the differential-integral equation. This means that the S^* functions can be computed once and for all, for all orders of scattering of interest, from which S functions for any λ can be determined. The relation (5.12) implies that, for $\lambda = 0.1$, about two scatterings are of interest while, for $\lambda = 0.5$, perhaps 6 or 7 scatterings should be considered. As we saw in Chapter 2, the conservative case of $\lambda = 1$ is quite different from all the rest. Thus, for conservative scattering, the exact equations of Chapter 2 should be used. Eq. (5.10) is to be solved subject to the initial condition

$$S(0; v, u) = 0. \tag{5.13}$$

As expected, this function S is the same as the function S of previous chapters.

5.2.2 Finite order transmission functions

In a manner similar to the above, differentiating eq. (5.4) with respect to t, passing to the limit $t = 0$, and making use of eq. (5.1), after some minor manipulations, we obtain the required invariant imbedding equation for the transmission function of order n,

$$\frac{\partial T}{\partial x}(n, x; v, u) + \frac{T}{v}(n, x; v, u) \tag{5.14}$$

$$= \lambda[e^{-x/u}\delta(n, 1) + \frac{1}{2}\int_0^1 T(n-1, x; v, u)\frac{dw}{w}$$

$$+\frac{e^{-x/u}}{2}\int_0^1 S(n-1, x; u, w)\frac{dw}{w}$$

$$+\frac{1}{4}\sum_{m=1}^{n-1}\int_0^1 S(n-m, x; v, w)\frac{dw}{w}$$

$$\times \int_0^1 T(m-1, x; w', u)\frac{dw'}{w'}],$$

together with the initial conditions

$$T(n, 0; v, u) = 0 \quad \text{and} \quad T(0, x; v, u) = 0. \tag{5.15}$$

As stated earlier, this equation can also be easily written down as is usual in the invariant imbedding method, by relating the transmitted intensity from a slab of thickness x to that from a slab of thickness $x+\Delta$, in the case of each order of scattering. Putting $n = 1$ in eq. (5.14), the transmission function of first order reduces to

$$T(1, x; v, u) = \lambda vu[\exp(-x/v) - \exp(-x/u)]/(v - u) \quad \text{for } v \neq u, \tag{5.16}$$

and

$$T(1, x; v, u) = \lambda xe^{-x/v} \quad \text{for } v = u. \tag{5.17}$$

By using eqs. (5.5), (5.8), (5.14) and (5.16), we get iteratively the required transmission function of the nth order. Eq. (5.14) can be used for numerical computation by high-speed computers. In a manner similar to the case of scattering function, we also have the reciprocity principle for the transmission function

$$T(n, x; v, u) = T(n, x; u, v). \tag{5.18}$$

Putting

$$T(n, x; v, u) = \lambda^n T^*(n, x; v, u), \tag{5.19}$$

and writing

$$T(x; v, u) = \sum_{n=1}^{\infty} T(n, x; v, u), \tag{5.20}$$

starting with eq. (5.14), we obtain the functional equation for the total transmission function as below

$$\frac{\partial T}{\partial x}(x; v, u) + \frac{T}{v}(x; v, u) \tag{5.21}$$

$$= \lambda[e^{-x/u} + \frac{1}{2}\int_0^1 T(x; w, u)\frac{dw}{w} + \frac{e^{-x/u}}{2}\int_0^1 S(x; u, w)\frac{dw}{w}$$

$$+\frac{1}{4}\int_0^1 S(x; v, w)\frac{dw}{w}\int_0^1 T(x; w', u)\frac{dw'}{w'}],$$

together with the initial condition

$$T(0; v, u) = 0. \tag{5.22}$$

These equations for T agree with those obtained in Chapter 2.

5.3 The Auxiliary Equation and its Solution

In the theory of radiative transfer, it is usual to define a source function $J(t, v, x)$ ($0 \leq t \leq x$, $0 < v \leq 1$) at any level t in the medium of optical thickness x, which represents the total rate of production of photons per unit volume per unit solid angle at level t above the level. The source function is related to the scattering function of radiation emerging from the top and, furthermore, to the transmission function of radiation escaping from the bottom in a simple fashion, namely as a truncated Laplace transform. The equation governing the source function is called the auxiliary equation, which coincides with the Milne integral equation with an exponential forcing function. This is discussed in Appendix B.

In this section we shall define the J-function corresponding to finite order n ($n \geq 1$) of scattering and obtain the integro-differential recurrence relation for the $S(n, x; v, u)$ and $T(n, x; v, u)$ functions (see eqs. (5.5) and (5.14)) in an alternate derivation that makes use of the Milne integral equation for the $J(n; t, v, x)$ function. We utilize a result analogous to the Bellman-Krein formula for the resolvent kernel for a Fredholm integral equation. See Appendix B. This results directly in the invariant imbedding equations governing the $S(n; x; v, u)$ and $T(n, x; v, u)$ functions.

The auxiliary equation for the source function $J(n; t, v, x)$ corresponding to the nth order scattered radiation takes the form

$$J(n; t, v, x) = (\lambda F/4)e^{-(x-t)/v}\delta(n, 1) \tag{5.23}$$
$$+(\lambda/2) \int_0^x E_1(|t - y|)J(n - 1; y, v, x)dy,$$

where $0 < u \leq 1$, $E_1(s)$ is the first exponential integral for positive real argument $0 \leq s \leq x$

$$E_1(s) = \int_0^1 e^{-s/w} dw/w. \tag{5.24}$$

With the aid of the Fredholm resolvent $K(n; t, y, x)$, eq. (5.23) is expressed in the form

$$J(n; t, v, x) = (\lambda F/4)\left[e^{-(x-t)/v}\delta(n, 1) \right. \tag{5.25}$$
$$\left. + \int_0^x K(n - 1; t, y, x)e^{-(x-y)/v}dy\right]$$

where $n \geq 1$, $0 \leq t$, $y \leq x$, and the resolvent satisfies the recurrence relations

$$K(n;t,y,x) = (\lambda/2)E_1(|t-y|)\delta(n,1) \tag{5.26}$$
$$+(\lambda/2)\int_0^x E_1(|t-z|)K(n-1;z,y,x)dz,$$

$$K(n;t,y,x) = (\lambda/2)E_1(|t-y|)\delta(n,1) \tag{5.27}$$
$$+(\lambda/2)\int_0^x K(n-1;t,z,x)E_1(|z-y|)dz.$$

Because of the displacement character of the kernel E_1-function, the K-function is also symmetric with respect to t and y.

On differentiating eq. (5.25) with respect to x, we have

$$\frac{\partial J}{\partial x}(n;t,v,x) = -\frac{J}{v}(n;t,v,x) + \frac{\lambda F}{4}P(n-1;t,x) \tag{5.28}$$
$$+\frac{\lambda F}{4}\int_0^x \frac{\partial K}{\partial x}(n-1;t,y,x)e^{-(x-y)/v}dy,$$

where

$$P(n;t,x) = K(n;t,x,x) = K(n;x,t,x). \tag{5.29}$$

Differentiation of the iterative kernel with respect to x leads to an expression similar to that of Bellman-Krein's formula as explained in Appendix B:

$$\frac{\partial K}{\partial x}(n;t,y,x) = \sum_{m=1}^{n-1} P(m;t,x)P(n-m;y,x). \tag{5.30}$$

On substituting eq. (5.30) into eq. (5.28) and recalling eq. (5.23), after some minor rearrangement of terms, we obtain

$$\frac{\partial J}{\partial x}(n;t,v,x) = -\frac{J}{v}(n;t,v,x) \tag{5.31}$$
$$+\frac{\lambda F}{4}\left\{ P(n-1;t,x) + \sum_{m=1}^{n-2} P(m;t,x) \right.$$
$$\times \left. \int_0^x P(n-m-1;y,x)e^{-(x-y)/v}dy \right\}$$
$$= -\frac{J}{v}(n;t,v,x) + \sum_{m=1}^{n-1} P(m;t,x)J(n-m;x,v,x),$$

where

$$P(m;t,x) \;=\; \frac{\lambda}{2}\int_0^1 e^{-(x-t)/w}\,\frac{dw}{w}\,\delta(m,1) \tag{5.32}$$

$$+\frac{\lambda}{2}\int_0^x\int_0^1 K(m-1;t,z,x)e^{-(x-z)/w}\,dz\,\frac{dw}{w}$$

$$=\;\frac{2}{F}\int_0^1 J(m;t,w,x)\,\frac{dw}{w},$$

by changing the order of integrations with respect to z and w in the second term on the right-hand side.

Let the scattering and transmission functions of the n-times scattered radiation be expressed in terms of the source function $J(n;t,v,x)$ of nth order by

$$S(n,x;v,u) = (4/F)\int_0^x J(n;t,v,x)e^{-(x-t)/u}\,dt, \tag{5.33}$$

$$T(n,x;v,u) = (4/F)\int_0^x J(n;t,v,x)e^{-t/u}\,dt. \tag{5.34}$$

On differentiating eq. (5.33) with respect to x, we have

$$\frac{\partial S}{\partial x}(n,x;v,u) \;=\; -\frac{S}{u}(n,x;v,u)+\frac{4}{F}J(n;x,v,x) \tag{5.35}$$

$$+\frac{4}{F}\int_0^x \frac{\partial J}{\partial x}(n;t,v,x)e^{-(x-t)/u}\,dt.$$

Recalling eq. (5.23), $J(n;x,v,x)$ is given by

$$J(n;x,v,x)=(\lambda F/4)\left[\delta(n,1)+(1/2)\int_0^1 S(n-1,x;v,w)\frac{dw}{w}\right]. \tag{5.36}$$

On inserting eqs. (5.31) and (5.36) into eq. (5.35), we get

$$\frac{\partial S}{\partial x}(n,x;v,u)+\left(\frac{1}{v}+\frac{1}{u}\right)S(n,x;v,u) \tag{5.37}$$

$$=\lambda\Big[\delta(n,1)+\frac{1}{2}\int_0^1 S(n-1,x;v,w)\frac{dw}{w}$$

$$+\frac{1}{2}\int_0^1 S(n-1,x;w,u)\frac{dw}{w}$$

$$+\frac{1}{4}\sum_{m=1}^{n-2}\int_0^1 S(m,x;w,u)\frac{dw}{w}$$

$$\times\int_0^1 S(n-m-1,x;v,w')\frac{dw'}{w'}\Big].$$

Similarly, differentiating eq. (5.34) with respect to x and recalling eqs. (5.31) and (5.32), we arrive at the following recurrence relation:

$$\frac{\partial T}{\partial x}(n, x; v, u) + \frac{T}{v}(n, x; v, u) \tag{5.38}$$

$$= \lambda \left[e^{-x/u} \delta(n, 1) + \frac{1}{2} \int_0^1 T(n-1, x; v, u) \frac{dw}{w} \right.$$

$$+ \frac{e^{-x/u}}{2} \int_0^1 S(n-1, x; u, w) \frac{dw}{w}$$

$$+ \frac{1}{4} \sum_{m=1}^{n-2} \int_0^1 S(n-m-1, x; v, w) \frac{dw}{w}$$

$$\left. \times \int_0^1 T(m, x; w', u) \frac{dw'}{w'} \right].$$

It should be mentioned that eqs. (5.37) and (5.38) reduce to eqs. (5.5) and (5.14), respectively, allowing for the reciprocity principle. It is relevant to point out that the invariant imbedding argument based on the particle-counting procedure permits us directly to write down eq. (5.31), and eq. (5.23) governing the multiple scattering processes can also be derived from the probabilistic aspects. The simultaneous usage of eqs. (5.29) through (5.32), together with the differential equations for P, leads to the determination of the scattering and transmission functions of finite order.

5.4 Cumulative Functions

Previously we dealt with specific intensity at any level, the scattering and transmission functions of the n-times scattered radiation. In this section we shall discuss these cumulative functions of the radiation which has undergone scatterings once, twice or up to n times during its passage through the target slab. In the limit as the cumulative order n of scatterings tends to infinity, these functions which we call cumulative functions yield directly the usual specific intensity, the scattering and transmission functions of multiply scattered radiation in a slab. Let us now focus our attention on the equations governing these cumulative functions $\bar{I}(n, t, v)$, $\bar{S}(n, x; v; u)$ and $\bar{T}(n, x; v, u)$.

Let us define $\bar{I}(n, t, v)$ as the specific intensity of radiation which has undergone scatterings up to and including n times ($n \geq 1$) within the medium, at any level t ($0 \leq t \leq x$) of the slab in a direction v ($0 < v \leq 1$) measured from the downward normal. The equation of transfer appropriate to this case takes the form

$$v \frac{dI}{dx}(n, t, v) = \bar{I}(n, t, v) - \frac{\lambda}{2} \int_{-1}^{+1} \bar{I}(n-1, t, w) dw \tag{5.39}$$

$$- \frac{\lambda F}{4} e^{-(x-t)/u},$$

together with the boundary conditions

$$\bar{I}(n, x, +v) = 0 \quad \text{and} \quad \bar{I}(n, 0, -v) = 0. \tag{5.40}$$

Eq. (5.39) can be readily obtained by the linear addition of the transfer equation for $I(m, t, v)$ (eq. (5.1)) over all values of m from $m = 1$ to $m = n$. In what follows, with the aid of an initial value method, we shall find the recurrence relations for the cumulative radiation field. The procedure does not refer to the recurrence relations given in the previous section.

As was stated earlier, corresponding to the source function $J(n; t, v, x)$ of the n times scattered radiation, we can consider the cumulative source function

$$M(n; t, v, x) = \sum_{m=1}^{n} J(m; t, v, x), \tag{5.41}$$

which satisfies the auxiliary equation

$$\begin{aligned} M(n; t, v, x) &= \frac{\lambda F}{4} e^{-(x-t)/v} + \frac{\lambda}{2} \\ &\times \int_0^x E_1(|t - y|) M(n - 1; y, v, x) dy. \end{aligned} \tag{5.42}$$

The resolvent kernel of this equation, which we call the cumulative resolvent

$$L(n; t, y, x) = \sum_{m=1}^{n} K(n; t, y, x), \tag{5.43}$$

is symmetric with respect to t and y. We can rewrite eq. (5.42) as below:

$$M(n; t, v, x) = \frac{\lambda F}{4} \left[e^{-(x-t)/v} + \int_0^x L(n - 1; t, y, x) e^{-(x-y)/v} dy \right]. \tag{5.44}$$

From eqs. (5.42) and (5.44) we find that the resolvent L satisfies the integral equation like recurrence relations

$$\begin{aligned} L(n; t, y, x) &= (\lambda/2) E_1(|t - y|) \\ &+ (\lambda/2) \int_0^x E_1(|t - z|) L(n - 1; z, y, x) dz, \end{aligned} \tag{5.45}$$

$$\begin{aligned} L(n; t, y, x) &= (\lambda/2) E_1(|t - y|) \\ &+ (\lambda/2) \int_0^x L(n - 1; t, z, x) E_1(|z - y|) dz. \end{aligned} \tag{5.46}$$

Introducing the function $Q(m; t, x)$,

$$Q(m; t, x) = L(m; t, x, x), \tag{5.47}$$

we have

$$\frac{\partial L^1}{\partial x}(n;t,y,x) \;=\; \sum_{m=1}^{n-1} Q(m;t,x)Q(n-m;y,x) \tag{5.48}$$

$$-\sum_{m=1}^{n-2} Q(m;t,x)Q(n-m-1;y,x),$$

where

$$Q(m;t,x) \;=\; \frac{\lambda}{2}\int_0^1 e^{-(x-t)/w}\frac{dw}{w} \tag{5.49}$$

$$+\frac{\lambda}{2}\int_0^x\int_0^1 L(m-1;t,z,x)e^{-(x-z)/w}dz\frac{dw}{w}$$

$$=\; \frac{2}{F}\int_0^1 M(m;t,w,x)\frac{dw}{w}.$$

In a manner similar to the case of eq. (5.32), the order of integrations is changed in the second term on the right-hand side. On differentiating eq. (5.44) with respect to x, we have

$$\frac{\partial M}{\partial x}(n;t,v,x) \;=\; -\frac{M}{v}(n;t,v,x) + \frac{\lambda F}{4}\Bigg[Q(n-1;t,x) \tag{5.50}$$

$$+\int_0^x \frac{\partial L}{\partial x}(n-1;t,z,x)e^{-(x-z)/v}dz\Bigg].$$

On making use of eqs. (5.47) and (5.48), eq. (5.50) becomes

$$\frac{\partial M}{\partial x}(n;t,v,x) \;=\; -\frac{M}{v}(n;t,v,x) + \frac{\lambda F}{4}\Bigg[Q(n-1;t,x) \tag{5.51}$$

$$+\sum_{m=1}^{n-2} Q(m;t,x)\int_0^1 Q(n-m-1;z,x)e^{-(x-z)/v}$$

$$-\sum_{m=1}^{n-3} Q(m;t,x)\int_0^1 Q(n-m-2;z,x)e^{-(x-z)/v}\Bigg].$$

Recalling eq. (5.44), eq. (5.51) reduces to

$$\frac{\partial M}{\partial x}(n;t,v,x) \;=\; -\frac{M}{v}(n;t,v,x) \tag{5.52}$$

$$+\sum_{m=1}^{n-1} Q(m;t,x)M(n-m;x,v,x)$$

$$-\sum_{m=1}^{n-2} Q(m;t,x)M(n-m-1;x,v,x).$$

Equation (5.52) is the required integro-differential type of the recurrence relation for the M-function.

Let us define the cumulative scattering and transmission functions as follows:

$$\bar{S}(n, x; v, u) = (4/F) \int_0^x M(n; t, v, x)e^{-(x-t)/u}\,dt, \tag{5.53}$$

$$\bar{T}(n, x; v, u) = (4/F) \int_0^x M(n; t, v, x)e^{-t/u}\,dt, \tag{5.54}$$

where the relationship between the cumulative and the finite order functions are given by

$$\bar{S}(n, x; v, u) = \sum_{m=1}^{n} S(m, x; v, u), \tag{5.55}$$

$$\bar{T}(n, x; v, u) = \sum_{m=1}^{n} T(m, x; v, u). \tag{5.56}$$

Combining eqs. (5.49) and (5.53), we have

$$\int_0^x Q(n; t, x)e^{-(x-t)/v}\,dt = \frac{1}{2}\int_0^1 \bar{S}(n, x; w, v)(dw/w), \tag{5.57}$$

and also from eq. (5.54) for $t = x$ we get

$$\begin{aligned}
M(n; x, v, x) &= \frac{\lambda F}{4}\left[1 + \int_0^x Q(n-1; z, x)e^{-(x-z)/v}\,dz\right] \tag{5.58}\\
&= \frac{\lambda F}{4}\left[1 + \frac{1}{2}\int_0^1 \bar{S}(n-1, x; w, v)\frac{dw}{w}\right].
\end{aligned}$$

Differentiation of eq. (5.53) with respect to x yields

$$\begin{aligned}
\frac{\partial \bar{S}}{\partial x}(n, x; v, u) &= -\frac{\bar{S}}{u}(n, x; v, u) + \frac{4}{F}M(n; x, v, x) \tag{5.59}\\
&\quad + \frac{4}{F}\int_0^x \frac{\partial M}{\partial x}(n; t, v, x)e^{-(x-t)/u}\,dt.
\end{aligned}$$

On inserting eqs. (5.52) and (5.58) into eq. (5.59), after some minor manipulations, we obtain the system of integro-differential recurrence relations

$$\frac{\partial \bar{S}}{\partial x}(n, x; v, u) + \left(\frac{1}{v} + \frac{1}{u}\right)\bar{S}(n, x; v, u) \tag{5.60}$$

$$\begin{aligned}
&= \lambda\left[1 + \frac{1}{2}\int_0^1 \bar{S}(n-1, x; w, v)\frac{dw}{w} + \frac{1}{2}\int_0^1 \bar{S}(n-1, x; w, u)\frac{dw}{w}\right.\\
&\quad + \frac{1}{4}\sum_{m=1}^{n-2}\int_0^1 \bar{S}(m, x; w, v)\frac{dw}{w}\int_0^1 \bar{S}(n-m-1, x; w', u)\frac{dw'}{w'}\\
&\quad \left. - \frac{1}{4}\sum_{m=1}^{n-3}\int_0^1 \bar{S}(m, x; w, v)\frac{dw}{w}\int_0^1 \bar{S}(n-m-2, x; w', u)\frac{dw'}{w'}\right].
\end{aligned}$$

The initial conditions are $\bar{S}(1, x; v, u)$ coincides with $S(1, x; v, u)$, given by eq. (5.8). It is readily proved that, summing eq. (5.37) over n and putting

$$\bar{S}(n, x; v, u) = \sum_{m=1}^{n} S(m, x; v, u), \tag{5.61}$$

the $\bar{S}(n, x; v, u)$ satisfies eq. (5.60), where $S(m, x; v, u)$ is the mth order scattering function.

Similarly, differentiating eq. (5.54) with respect to x, we get

$$\frac{\partial \bar{T}}{\partial x}(n, x; v, u) = \frac{4}{F} M(n; x, v, x) e^{-x/u} \tag{5.62}$$
$$+ \frac{4}{F} \int_0^x \frac{\partial M}{\partial x}(n; t, v, x) e^{-t/u} dt.$$

Recalling eqs. (5.52) and (5.58), from eq. (5.62) we have

$$\frac{\partial \bar{T}}{\partial x}(n, x; v, u) = \lambda\left[1 + \frac{1}{2}\int_0^1 \bar{S}(n-1, x; w, v)\frac{dw}{w}\right]e^{-x/u} \tag{5.63}$$
$$+ \frac{4}{F}\int_0^x \left[-\frac{M}{v}(n; t, v, x)\right.$$
$$+ \sum_{m=1}^{n-1} Q(m; t, x) M(n-m; x, v, x)$$
$$\left. - \sum_{m=1}^{n-2} Q(m; t, x) M(n-m-1; x, v, x)\right]e^{-t/u} dt.$$

On making use of eqs. (5.49), (5.54), and (5.58), we obtain the system of recurrence relations as follows:

$$\frac{\partial \bar{T}}{\partial x}(n, x; v, u) + \frac{\bar{T}}{v}(n, x; v, u) \tag{5.64}$$
$$= \lambda\left[e^{-x/u} + \frac{1}{2}\int_0^1 \bar{T}(n-1, x; w, u)\frac{dw}{w} + \frac{e^{-x/u}}{2}\int_0^1 \bar{S}(n-1, x; w, v)\frac{dw}{w}\right.$$
$$+ \frac{1}{4}\sum_{m=1}^{n-2}\int_0^1 \bar{S}(n-m-1, x; w, v)\frac{dw}{w}\int_0^1 \bar{T}(m, x; w, u)\frac{dw}{w}$$
$$\left. - \frac{1}{4}\sum_{m=1}^{n-3}\int_0^1 \bar{S}(n-m-2, x; w, v)\frac{dw}{w}\int_0^1 \bar{T}(m, x; w, u)\frac{dw}{w}\right],$$

with initial conditions, where $n \geq 1$, $\bar{T}(1, x; v, n)$ is equal to $T(1, x; v, u)$ given by eqs. (5.16) and (5.17). It should be mentioned that

$$\bar{T}(n, x; v, u) = \sum_{m=1}^{n} T(m, x; v, u). \tag{5.65}$$

Eqs. (5.60) and (5.64) can be used for numerical computation by high-speed computers, if n is small.

For more details and related analyses, see reference [1]–[9].

5.5 Discussion

In this chapter we have established the scattering functions $S(n, x, v, u)$ and transmission functions $T(n, x, v, u)$ of n-th order scattering and for the cumulative functions as well. Even though no computational results are presented here, those functions are well suited for numerical treatment by modern high speed computers similar to those discussed in Chapters 2 and 3. Our approach also provides physical meaning to the problem by considering n-th order of scattering, $n = 1, 2, 3, \ldots$. Of course we obtain the exact solution as $n \to \infty$.

By determining the cumulative functions $\bar{S}$ and $\bar{T}$ comparing them with the exact S and T, we can learn how many scatterings contribute to the diffuse intensity fields.

References

1. R. Bellman, S. Ueno and R. Vasudevan, "Invariant Imbedding and Radiative Dosimetry: I. Finite Order Scattering and Transmission Function," *Math. Bioscience*, Vol. 14, 1972, pp. 235-254.

2. R. Bellman, S. Ueno and R. Vasudevan, "Invariant Imbedding and Dosimetry: II. Integral Recurrence Relations for the Finite Order and Transmission Functions," *Math. Bioscience*, Vol. 15, 1972, pp. 153–162.

3. S. Ueno and A. P. Wang, "Scattering and Transmission Functions of Radiation by Finite Atmospheres with Reflection Surfaces," *Astrophys. Space Science*, Vol. 23, 1973, pp. 205–220.

4. R. Bellman, S. Ueno and R. Vasudevan, "Invariant Imbedding and Radiation Dosimetry: VII. Finite Order Scattering and Transmission Functions of the Two Radiation Approximations in a Target Slab," *Math. Biosciences*, Vol. 18, 1973, pp. 255–268.

5. S. Ueno, "Scattering and Transmission Matrices of Partially Polarized Radiation in a Rayleigh Atmosphere Bounded by a Specular Reflector," *Bull. Amer. Astr. Soc.*, Vol. 5, 1973, p. 304.

6. S. Ueno and S. Mukai, "The Contour of Absorption Lines in a Moving Atmosphere," in "Special Report, No. 1," *Research Institute for Information Science, Kanazawa Institute of Technology*, pp. 1–15.

7. S. Ueno and A. P. Wang, "Finite-Order Scattering and Transmission Functions in Chandrasekhar's Planetary Problem," *Bull. Amer. Astr. Soc.*, Vol. 8, 1977, pp. 472–477.

8. K. Kawabata and S. Ueno, "The First Three Orders of Scattering in Vertically Inhomogeneous Scattering-Absorbing Media," *Astrophysics and Space Science*, Vol. 150, 1988, pp. 327–344.

9. S. Ueno and A. P. Wang, "Order-of-Scattering Theory in Radiation Field," *Computer Math. Applic.*, Vol. 27, No. 9, 1994, pp. 169–173.

6. Scattering Matrix

In previous chapters, we developed the invariant imbedding technique for multiple scattering in a vertically stratified plane parallel medium. Many advantages of this technique have been discussed. The purpose of this chapter is to extend invariant imbedding techniques by introducing scattering matrix analysis. The scattering matrix relates inputs to outputs. Such an extension gives us a physical understanding of complex multiple scattering. This approach provides a more solid mathematical structure. This also provides us with a new tool to solve more complicated problems such as time-dependent radiative transfer. Scattering matrix analysis lays the foundation of computational methods, several as presented in this chapter. It also leads us naturally to discuss and solve the inverse problems which are used in other chapters.

6.1 Introduction

The scattering matrix relates inputs and outputs at the boundaries of a plane parallel layer. It involves the transmission and reflection operators in the forward and backward directions. The assembly of two scattering matrices, called the star-product, is based on the principle of invariant imbedding discussed in previous chapters. Our results include some known properties such as Stokes commutativity and Ambarzumian's principle. The transport equations, which govern the multiple scattering processes, are constructed. As in the classic physics and mathematical semi-group theory, these equations relate the transmission and reflection operators to a single generator. Because of this general setting, the scattering matrix can be easily extended.

Along with the analysis, we take computation into consideration. We develop the diagonalization of a scattering matrix, the n-th term approximate solution. We also develop a discrete model designed for use in the standard sweep method computation to solve the complicated time-dependent case.

Since a large amount of information is assembled in this chapter, some of the details are not presented here. The interested reader can find details in the references at the end of the chapter [1]–[12]. The important points are that we provide the reader the mathematical structure and computational methods to solve a class of radiative transfer problems.

6.2 The Scattering Matrix

Let us consider a layer of atmospheric plane parallel medium extending from depth $x = x_1$ to $x = x_2$. Let $I^+(x_1)$ be the input incident radiation at the top and simultaneously $I(x_2)$ at the bottom, then there are output radiations $I^+(x_2)$ and $I^-(x_1)$. As shown in Figure 6.1, we use I^+ as the downward direction and I^- as the upward direction. The relations between inputs and outputs are governed by the scattering process so that the transmissions and reflections are taking place. For example, if only a single input $I^+(x_1)$ is incident at the top, then we have the output $I^+(x_2)$ given by the transmission operator, t, at the bottom and output $I^-(x_1)$ given by the reflection operator, r, at the top, i.e.,

$$I^+(x_2) = t \cdot I^+(x_1) \quad \text{and} \quad I^-(x_1) = r \cdot I^+(x_1). \tag{6.1a}$$

Since the system is linear, the transmission and reflection operators are linear. Integral operators are commonly used in radiative transfer. Likewise we use transmission τ and reflection ρ when a single input $I^-(x_2)$ is incident at the bottom, i.e.,

$$I^-(x_1) = \tau \cdot I^-(x_2) \quad \text{and} \quad I^+(x_2) = \rho \cdot I^-(x_2). \tag{6.1b}$$

Inputs and outputs in radiative transfer are intensities. For example, $I^+(x_1) = I(x_1, \theta, \phi)$ is the intensity in the downward direction, at depth x_1 in the direction (θ, ϕ). Figure 6.2 shows that two inputs $I^+(x_1)$ and $I^-(x_2)$ are incident at the top and at the bottom, respectively, of a parallel layer. Under the multiple scattering process, two outputs $I^-(x_1)$ and $I^+(x_2)$ are produced. The linear system allows us to combine outputs by addition when there are two simultaneous inputs. By combining outputs $I^-(x)$ and $I^+(x)$ in eq. (6.1a) and (6.1b), we have

$$I^+(x_2) = t \cdot I^+(x_1) + \rho \cdot I^-(x_2) \tag{6.2a}$$

and

$$I^-(x_1) = r \cdot I^+(x_1) + \tau \cdot I^-(x_2). \tag{6.2b}$$

Equations (6.2a) and (6.2b) can be combined into a single matrix equation,

$$\begin{pmatrix} I^+(x_2) \\ I^-(x_1) \end{pmatrix} = \begin{pmatrix} t & \rho \\ r & \tau \end{pmatrix} \begin{pmatrix} I^+(x_1) \\ I^-(x_2) \end{pmatrix} \tag{6.3}$$

$$\underbrace{\hphantom{\begin{pmatrix} I^+(x_2) \\ I^-(x_1) \end{pmatrix}}}_{\text{outputs}} \qquad\qquad \underbrace{\hphantom{\begin{pmatrix} I^+(x_1) \\ I^-(x_2) \end{pmatrix}}}_{\text{inputs}}$$

Let us define the 2×2 matrix of operators

$$\begin{pmatrix} t & \rho \\ r & \tau \end{pmatrix} = \mathfrak{S}(x_1, x_2), \tag{6.4}$$

the *scattering matrix*. Eq. (6.3) indicates the scattering matrix controls the whole scattering process. It maps inputs to outputs. The partition of $\mathfrak{S}$ into

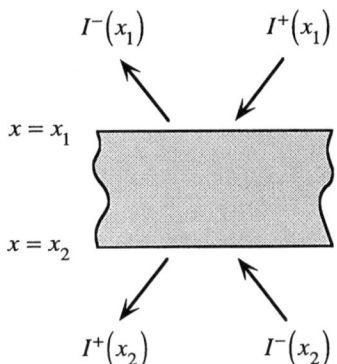

Figure 6.1. Inputs & outputs.

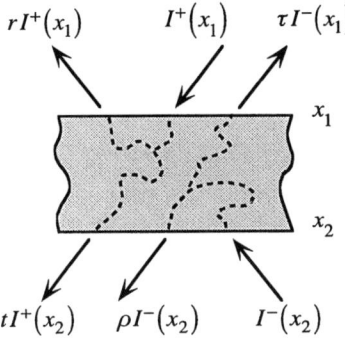

Figure 6.2. Transmissions & reflections

four operators t, τ, ρ and r enables us to give physical meaning and the scattering matrix is the basic element of the following mathematical analysis.

If an additional plane parallel layer is immediately below the above layer for depth extending from $x = x_2$ to $x = x_3$. Corresponding to inputs $I^+(x_1)$ and $I^-(x_3)$ there is multiple scattering taking place within those two layers. Then $\mathfrak{S}_1$ is the scattering matrix for the added layer. The subscript "1" refers to the added layer below. As the scattering matrix $\mathfrak{S}$, the scattering matrix

$$\mathfrak{S}(x_2, x_3) = \begin{pmatrix} t_1 & \rho_1 \\ r_1 & \tau_1 \end{pmatrix} \tag{6.5}$$

relates inputs to outputs at the boundaries x_2 and x_3, i.e.,

$$\underbrace{\begin{pmatrix} I^+(x_3) \\ I^-(x_2) \end{pmatrix}}_{\text{outputs}} = \begin{pmatrix} t_1 & \rho_1 \\ r_1 & \tau_1 \end{pmatrix} \underbrace{\begin{pmatrix} I^+(x_2) \\ I^-(x_3) \end{pmatrix}}_{\text{inputs}} \tag{6.6}$$

To establish the scattering matrix, $\mathfrak{S}(x_1, x_3)$, for the combined layers, i.e.,

$$\begin{pmatrix} I^+(x_3) \\ I^-(x_1) \end{pmatrix} = \mathfrak{S}(x_1, x_3) \begin{pmatrix} I^+(x_1) \\ I^-(x_3) \end{pmatrix}, \tag{6.7}$$

We can solve $I^+(x_3)$ and $I^-(x_1)$ in terms of $I^+(x_1)$ and $I^-(x_3)$ by eqs. (6.3) and (6.6). The result is

$$\mathfrak{S} * \mathfrak{S}_1 = \begin{pmatrix} t & \rho \\ r & \tau \end{pmatrix} * \begin{pmatrix} t_1 & \rho_1 \\ r_1 & \tau_1 \end{pmatrix}$$

$$= \begin{pmatrix} t_1(E - \rho r_1)^{-1}t & \rho_1 + t_1\rho(E - r_1\rho)^{-1}\tau_1 \\ r + \tau r_1(E - \rho r_1)^{-1}t & \tau(E - r_1\rho)^{-1}\tau_1 \end{pmatrix}, \quad (6.8)$$

where E is the identity operator.

For convenience, eq. (6.8) is expressed as

$$\mathfrak{S}(x_1, x_3) = \mathfrak{S}(x_1, x_2) * \mathfrak{S}(x_2, x_3). \quad (6.9)$$

As indicated by (6.7) the scattering matrix $\mathfrak{S}(x_1, x_3)$ controls the overall reflections and transmissions of a combined two-layered atmosphere. This associative operator, called the *-product* or *star product* [1], specifies the algebraic structure of all scattering processes to which the *invariant imbedding principle* [1], [2] applies. In the context of the star product, the invariant imbedding principle states that the scattering matrices must be invariant under the star product, i.e., when two layers are combined properties of each individual scattering matrix are unchanged. We can consider the invariance governing the law of diffuse reflection by a semi-infinite atmosphere first formulated by Ambarzumian [3] as a special case.

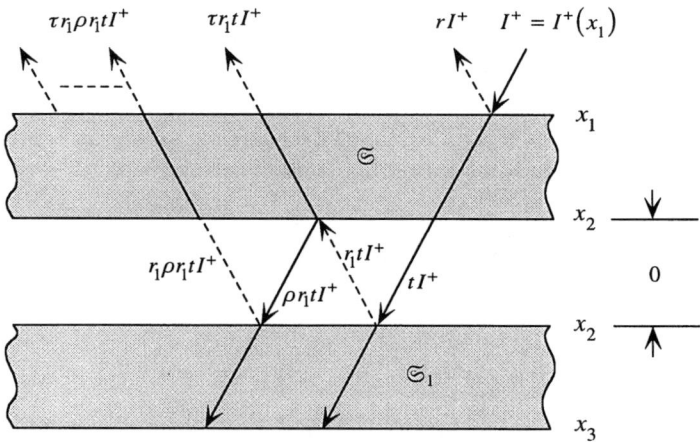

Figure 6.3. Multiple scattering

To illustrate the physical meaning of the star product, let us consider the total reflection of two combined layers, associated with scattering matrices $\mathfrak{S}$ and $\mathfrak{S}_1$, with input $I^+(x_1)$ on the top. In Fig. 6.3 we show that part of the input is immediately reflected upward denoted by rI^+ and part is transmitted downward denoted by tI^+. There is multiple scattering taking place between two layers. Part of tI^+ is reflected by the lower layer denoted by r_1tI^+. Again part of r_1tI^+ is reflected downward by ρr_1tI^+ and upward by $r_1\rho r_1tI^+$. And it is transmitted upward by the amount at x_1, by $\tau_1 r_1 \rho r_1 t$. This process continues. The total intensity reflected which forms the composite layer is

$$I^-(x_1) = [r + \tau r_1 t + \tau r_1 \rho r_1 t + \cdots + \tau r_1 (\rho r_1)^n t + \cdots]I^+(x_1) . \qquad (6.10)$$

If the series converges under the super norm $\|\ \ \|$, then the righthand side of eq. (6.10) can be expressed as $[r + \tau r_1(E - \rho r_1)^{-1}t]I^+(x_1)$ where E is the identity operator. We introduce the total reflection operator

$$Q(k) \overset{\Delta}{=} r + \tau k(E - \rho k)^{-1}t \qquad (6.11)$$

$Q(k)$ corresponding to the terminating reflection k in terms of operators t, τ, ρ, r for the layer. So that eq. (6.10) becomes

$$I^-(x_1) = Q(r_1)I^+(x_1) \qquad (6.12)$$

the total reflection $I^+(x_1)$ at the top of the combined layers is governed by the total reflector operator $Q(r_1)$. The operator $Q(r_1)$ is only dependent on r_1 of the added layer. The same result holds if the added layer is replaced by a reflector with reflection operator r_1. In fig. 6.4, we show a layer associated with scattering matrix $\mathfrak{S}$ is bounded below by a reflector with reflection operator k, then the total reflection operator is $Q(k)$, as in eq. (6.12) r_1 is replaced by k.

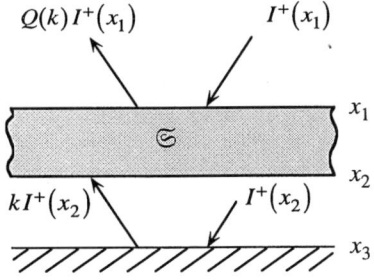

Figure 6.4. Total reflection

To add insight to $(E - \rho r_1)^{-1}$ and $(E - r_1\rho)^{-1}$, we should carry out our analyses without use of the inverse. Let p and q be the resolvents of (ρ, r_1) and (r_1, ρ), respectively, i.e., they are solutions of $p = E + p\rho r_1$ and $q = E + qr_1\rho$.

The existence of these resolvents guarantees that $\mathfrak{S} * \mathfrak{S}_1$ is well-defined. It is easy to show that the existence of $(E - \rho r_1)^{-1}$ implies the existence of $(E - r_1\rho)^{-1}$ and vice versa. If p exists, let $q = E + r_1 p\rho$, then

$$q = E + r_1(E + p\rho r_1)p = E + (E + r_1 p\rho)r_1\rho = E + qr_1\rho . \tag{6.13}$$

Hence, q is the resolvent of (r_1, ρ). The converse can be proved similarly by letting $p = E + \rho q r_1$.

The condition $r_1\rho = E$ (or $\rho r_1 = E$) means that the star product of two media does not exist, since the star product involves the operator $(E - r_1\rho)^{-1}$. On the other hand, if $||\rho r_1|| < 1$, then the star product is well-defined. The condition $r_1\rho = E$ means that we have a resonant condition in physics. For radiative transfer in a plane parallel atmosphere, if the one layer is *dissipative*, $||\mathfrak{S}|| < 1$, and the other is *lossless*, $||\mathfrak{S}|| = 1$, or dissipative, then the star-product is always well-defined. Dissipative and lossless conditions are easily satisfied if there is a loss of radiative energy, as for example if part of the radiation energy is turned into thermal energy.

6.3 The Homogeneous Medium

If the medium is *homogeneous*, in the sense that its scattering properties depend on the thickness only, i.e., $\mathfrak{S}(x_1, x_2) = \mathfrak{S}(x_2 - x_1)$. Then eq. (6.9) can be written

$$\mathfrak{S}(x_3 - x_1) = \mathfrak{S}(x_2 - x_1) * \mathfrak{S}(x_3 - x_2) . \tag{6.14}$$

Hence, if a layer of thickness x_1 is adjacent to a layer of thickness x_2, we get the *Stokes commutativity* relation

$$\mathfrak{S}(x_1) * \mathfrak{S}(x_2) = \mathfrak{S}(x_2) * \mathfrak{S}(x_1) . \tag{6.15}$$

Suppose that the thickness is infinite or semi-infinite, so that

$$\lim_{x \to \infty} \mathfrak{S}(x) = \mathfrak{S}_\infty . \tag{6.16}$$

In this case it is physically reasonable to assume that

$$t_\infty = \tau_\infty = 0 \tag{6.17}$$

i.e. the medium is so thick that no radiation is transmitted. Use of eq. (6.9) and eq. (6.16) produces two relations, namely,

$$\tau_\infty = \tau + \tau r_\infty (E - \rho r_\infty)t^{-1}, \qquad \rho_\infty = \rho + t\rho_\infty (E - r\rho_\infty)^{-1}\tau . \tag{6.18}$$

These equations give the limiting reflections as a given layer is combined with an infinitely thick medium. Although two equations are involved in eq. (6.18), in reality there is only one, in that r_∞ and ρ_∞^{-1} both satisfy the *fixed-point* equation

$$Q(k) = k , \qquad (6.19)$$

where $Q(k)$ is as defined in eq. (6.11). Applying this fixed point property to homogeneous layers, we obtain

$$\mathfrak{S}(x) * \mathfrak{S}_\infty = \mathfrak{S}_\infty * \mathfrak{S}(x) = \mathfrak{S}_\infty . \qquad (6.20)$$

The above is known as the Ambarzumian invariance principle. This is a consequence of Stokes commutativity.

6.4 The Transport Equation

Let us consider a thin layer extending from x_2 to $x_2 + \Delta x$. It is reasonable to assume that the scattering matrix of a thin medium differs from the identity by an amount proportional to the thickness. Thus,

$$
\begin{aligned}
\mathfrak{S}(x_2, x_2 + \Delta x) &= \begin{pmatrix} E & 0 \\ 0 & E \end{pmatrix} + \begin{pmatrix} b & a \\ c & d \end{pmatrix} \Delta x + o(\Delta x) \qquad (6.21) \\
&= E + M \Delta x + o(\Delta x)
\end{aligned}
$$

where a, b, c and d are operators depending on x_2 and $o(\Delta x) \to 0$ as $\Delta x \to 0$. Using eq. (6.9), we obtain

$$\mathfrak{S}(x_1, x_2) * \mathfrak{S}(x_2, x_2 + \Delta x) = \mathfrak{S}(x_1, x_2 + \Delta x) . \qquad (6.22)$$

Although the star product is not distributive, we can directly compute

$$
\begin{aligned}
\begin{pmatrix} t & \rho \\ r & \tau \end{pmatrix} &* \begin{pmatrix} E + b\Delta x & a\Delta x \\ c\Delta x & E + d\Delta x \end{pmatrix} \qquad (6.23) \\
&= \begin{pmatrix} t + (b + \rho c)t\Delta x & \rho + (a + b\rho + \rho d + \rho c\rho)\Delta x \\ r + \tau c t\Delta x & \tau + \tau(b_2 - c\rho)\Delta x \end{pmatrix}
\end{aligned}
$$

Subtracting $\mathfrak{S}(x_1, x_2)$, dividing by Δx and taking the limit as $\Delta x \to 0$ gives the *(right) transport equation*

$$\frac{\partial \mathfrak{S}(x_1, x_2)}{\partial x_2} = \begin{pmatrix} (b + \rho c)t & a + b\rho + \rho d + \rho c\rho \\ \tau c t & \tau(d + c\rho) \end{pmatrix} . \qquad (6.24)$$

This procedure also applied to

$$\mathfrak{S}(x_1 - \Delta x, x_1) * \mathfrak{S}(x_1, x_2) = \mathfrak{S}(x_1 - \Delta x, x_2)$$

yields

$$-\frac{\partial \mathfrak{S}(x_1, x_2)}{\partial x_1} = \begin{pmatrix} t(b + ar) & ta\tau \\ c + dr + rb + rar & (d + ra)\tau \end{pmatrix} \qquad (6.25)$$

the *(left) transport equation.*

Eq. (6.21) can also be expressed as

$$\lim_{\Delta x \to 0} \frac{\mathfrak{S}(x_2, x_2 + \Delta x) - E}{\Delta x} = M(x_2) . \tag{6.26}$$

Mathematically, this equation establishes the relation between the scattering matrix, $\mathfrak{S}$, and the *generator* $M = \begin{pmatrix} a & b \\ c & d \end{pmatrix}$.

For a homogeneous medium, the addition of a thin medium at x_1 is equivalent to the addition of a thin medium at x_2, therefore,

$$\frac{\partial \mathfrak{S}(x_1, x_2)}{\partial x_2} = -\frac{\partial \mathfrak{S}(x_1, x_2)}{\partial x_1} , \tag{6.27}$$

and we deduce

$$\begin{array}{cc} (b + \rho c)t = t(b + ar) & a + b\rho + \rho d + \rho c\rho = ta\tau \\ \tau ct = c + rb + dr + rar & \tau(d + c\rho) = (d + ra)\tau . \end{array} \tag{6.28}$$

In the case where the medium has infinite thickness and there is some dissipation, then

$$\frac{\partial \rho(x_1, x_2)}{\partial x_2} \to 0 \quad \text{and} \quad \frac{\partial r(x_1, x_2)}{\partial x_1} \to 0 \tag{6.29}$$

as $x_1 - x_2 \to \infty$. An immediate consequence is that

$$ta\tau \to 0 \quad \text{and} \quad \tau ct \to 0 \tag{6.30}$$

for infinite homogeneous media. This is a weaker requirement than that assumed in the discussion before eq. (6.17).

In a *free space*, that is, a medium embedded in a vacuum, the initial value for equations of transfer is

$$\mathfrak{S} = \begin{pmatrix} E & 0 \\ 0 & E \end{pmatrix} \quad \text{for } x_1 = x_2 . \tag{6.31}$$

For a plane parallel atmosphere, the *generator* has the form

$$M = M(x, \mu, \varphi) = \begin{pmatrix} \frac{1}{\mu}\delta & 0 \\ 0 & -\delta\frac{1}{\mu} \end{pmatrix} e^{-\alpha x/\mu} \tag{6.32}$$

$$+ \frac{1}{4\pi\mu} \begin{pmatrix} \gamma^{++} & \gamma^{-+} \\ -\gamma^{+-} & -\gamma^{--} \end{pmatrix} .$$

The first term is the *specular part* and the second is the *diffuse part*, where δ, α and $\gamma^{\pm\pm}$ are, respectively, the Dirac delta operator, the volume attenuation and the normalized x-dependent phase operator. For example,

$$(\gamma^{-+}\rho) = \int_0^1 \int_0^{2\pi} p(x; -\mu, \varphi; \tilde{\mu}, \tilde{\varphi})\rho(\tilde{\mu}, \tilde{\varphi}; \mu_0, \varphi_0) \, d\tilde{\mu} \, d\tilde{\varphi}) , \tag{6.33}$$

where p is the normalized phase function. Likewise, the exact meanings of $\frac{1}{\mu}\delta$ and $\delta\frac{1}{\mu}$ are

$$\left(\frac{1}{\mu}\delta\rho\right)(-\mu,\varphi) = \int_0^1 \int_0^{2\pi} \frac{1}{\mu}\delta(\mu-\tilde{\mu})\delta(\phi-\tilde{\phi})\rho(\tilde{\mu},\tilde{\phi};\mu_0,\phi_0)\,d\tilde{\mu}\mu\tilde{\phi} \quad (6.34)$$

and

$$\left(\delta\frac{1}{\mu}\rho\right)(-\mu,\varphi) = \int_0^1 \int_0^{2\pi} \delta(\mu-\tilde{\mu})\delta(\phi-\tilde{\phi})\frac{1}{\mu_0}\rho(\mu,\phi;\mu_0,\phi_0)\,d\tilde{\mu}\,d\tilde{\phi}. \quad (6.35)$$

Upon substitution of the above integral operations into the reflection equations of (6.25), i.e.,

$$\frac{\partial\rho}{\partial x_2} = a + b\rho + \rho c + \rho c\rho, \quad \rho = \rho(x_1,x_2) \quad (6.36)$$

we obtain the integral (*right*) *equation of reflection*

$$\left(\frac{1}{\mu}+\frac{1}{\mu_0}\right)\frac{\partial\rho}{\partial x_2}(\mu,\varphi;-\mu_0,\varphi_0)$$

$$= \rho(\mu,\phi;-\mu_0,\phi_0)$$

$$+\frac{1}{4\pi}\int_0^1\int_0^{2\pi} p(\mu,\phi;-\mu',\phi')\rho(-\mu',\phi';-\mu_0,\phi_0)\frac{d\mu'}{\mu'}\,d\phi'$$

$$+\frac{1}{4\pi}\int_0^1\int_0^{2\pi} \rho(\mu,\phi;-\mu'',\phi'')p(-\mu'',\phi'';-\mu_0,\phi_0)\frac{d\mu''}{\mu''}\,d\phi'' \quad (6.37)$$

$$+\frac{1}{16\pi^2}\int_0^1\int_0^{2\pi}\int_0^1\int_0^{2\pi} \rho(\mu,\phi;\mu',\phi')p(-\mu',\phi';\mu'',\phi'')$$

$$\times\rho(\mu'',\phi'';-\mu_0,\phi_0)\frac{d\mu'}{\mu'}\,d\phi'\frac{d\mu''}{\mu''}\,d\phi'',$$

where $(\pm\mu_0,\phi_0)$ is the specific direction, as in Sect. 1.1, of the input intensity and $(\pm\mu,\phi)$ the specific direction of the output intensity. This equation has initial value

$$\rho(\mu,\phi;-\mu_0,\phi_0) = 0. \quad (6.38)$$

6.5 The Star-Semi-Group

The star product defined in the previous section is a generalization of a classic semi-group. More precisely, we say a family of two-parameter $\mathfrak{S}(x_1,x_2)$, $0 \leq x_1 \leq x_2 < \infty$, is said to be a *star semigroup* of a bounded linear operator if

  i) $\mathfrak{S}(x_1,x_1) = E$ (6.39)

  ii) $\mathfrak{S}(x_1,x_3) = \mathfrak{S}(x_1,x_2) * \mathfrak{S}(x_2,x_3)$ for $0 \leq x_1 \leq x_2 \leq x_3$.

It is called *strongly continuous* at x_1 if for each intensity II,

  iii) $||\mathfrak{S}(x_1,x_2)II - II|| \to 0$ as $x_2 \to x_1$, under the norm $||II||^2 = |I^+|^2 + |I^-|^2$, with $|I^\pm|^2$ as the L_2 norm.

We define

$$\hat{\mathfrak{S}} \triangleq \begin{pmatrix} t^{-1} & -t^{-1}\rho \\ rt^{-1} & \tau - rt^{-1}\rho \end{pmatrix} \tag{6.40}$$

In case t^{-1} is nonsingular and $\mathfrak{S}$ is homogeneous, then $\hat{\mathfrak{S}}(x_1, x_2) = \hat{\mathfrak{S}}(x_2 - x_1)$. The one-parameter family of $\hat{\mathfrak{S}}$ forms a *semigroup*,

   i) $\hat{\mathfrak{S}}(0) = E$ $\tag{6.41}$

   ii) $\hat{\mathfrak{S}}(x_1 + x_2) = \hat{\mathfrak{S}}(x_1) \cdot \hat{\mathfrak{S}}(x_2)$ for $0 \le x_1, x_2 < \infty$

where $\cdot$ is the usual matrix multiplication. The following properties can be obtained by use of the super norm:

   i)   the star product is strongly continuous

   ii) if $||\mathfrak{S}_1|| < 1$ and $||\mathfrak{S}_2|| \le 1$, then $||\mathfrak{S}_1 * \mathfrak{S}_2|| < \max(||\mathfrak{S}_1||, ||\mathfrak{S}_2||)$.

   iii) if $||\mathfrak{S}_1|| < 1$ and $||\mathfrak{S}_2|| = 1$, then $||\mathfrak{S}_1 * \mathfrak{S}_2|| = ||\mathfrak{S}_1||$

   iv) if $||\mathfrak{S}|| < 1$ and $||r|| < 1$, then $\tag{6.42}$

$$||Q(k)|| \le 1 \quad \text{whenever} \quad ||k|| \le 1 \tag{6.43}$$

where Q is as defined in eq. (6.11).

   v) If $\mathfrak{S}$ is self-adjoint, $||\mathfrak{S}|| < 1$ and $||r|| < 1$, see [7], then there exists a unitary matrix U such that

$$U * \mathfrak{S} * U^{-1} = D , \tag{6.44}$$

where D is a diagonal scattering matrix and $||\mathfrak{S}|| = ||D||$. We shall use such properties for approximate solutions in the next section.

In a plane parallel atmosphere, the operator $\mathfrak{S}$ is dependent on the pair of variables $(\pm\mu, \phi)$ and $(\pm\mu_0, \phi_0)$, i.e., the specific directions of output intensity and input intensity. The spatial inversion invariance discussed in Sect. 1.4 leads to

$$\gamma^{++} = \gamma^{--} \quad \text{and} \quad \gamma^{+-} = \gamma^{-+} \tag{6.45}$$

in eq. (6.32). It follows that $\mathfrak{S}$ is self-adjoint. Therefore, eq. (6.44) can be used in radiative transfer to diagonalize the scattering matrix.

6.6 The n Terms Solutions

In this section we present the solutions of a plane parallel layer extending from depth x_1 to x_2 with uniform inputs at the boundaries, $I^+(x_1)$ and $I^-(x_2)$. The results are in the form of Bremmer series. The first n terms of this series denote the intensities that are experienced physically exactly as the first n reflections in the medium.

The classic WKB method usually is used to compute the one dimensional solution of a wave through an inhomogeneous medium, such as the scalar Schrödinger equation. The WKB method is modified for this problem. Instead of considering the multiple scattering actually taking place, we approximate the process by considering only nth order scattering. For example, at first the input intensities $I^-(x_2)$ experience no reflection,

$$I_0^-(x) = \tau_0(x_1, x_2)I^-(x_2), \qquad x_1 \leq x \leq x_2 . \tag{6.46}$$

Let $I_1^+(h)$ denote the intensities resulting from exactly one reflection,

$$I_1^+(x) = \rho_1(x, x_2)I^-(x), \qquad x_1 \leq x \leq x_2 . \tag{6.47}$$

In computation, operators τ_0 and ρ_1 are approximated by generator M, see eqs. (6.32) and (6.31). This is exact, if we take the limit by using the product integer sign $\prod_x^{x_2}$, for details see [9]. For an approximation one may use the Gauss quadrature method in Chapter 2. In general, we have

$$I_{2n+2}^-(x) = \int_x^{x_2} \prod_\sigma^{x_2} \exp[-d(s)\,ds]c(\sigma)I_{2n+1}^+(\sigma)\,d\sigma \tag{6.48}$$

and

$$I_{2n+1}^+(x) = \int_{x_1}^x \prod_\sigma^x \exp[b(s)\,ds]a(\sigma)I_{2n}^-(\sigma)\,d\sigma , \tag{6.49}$$

where subscripts denote the exact order of reflections occurring in the process. We identify a, b, c and d as medium coefficients of a nonhomogeneous layer. $I_{2n+2}^-(x)$ and $I_{2n+1}^+(x)$ are internal intensities, moving in opposite directions. When $x = x_1$, then $I_{2n+2}^-(x_1)$ denotes the outputs at x_1 and when $x = x_2$, $I_{2n+1}^+(x_2)$ is the output at x_2. To take all multiple scattering into account,

$$I^-(x) = \sum_{n=0}^{\infty} I_{2n}^-(x) \quad \text{and} \quad I^+(x) = \sum_{n=0}^{\infty} I_{2n+1}^+(x) . \tag{6.50}$$

A Bremmer series solution is constructed. This is an exact solution. For approximate solutions one takes the first few terms. The convergence criteria have been established [9]. Usually the convergence is fast and depends on the thickness and the nature of the phase function.

6.7 The Discrete Case

For the discrete case the operators t, τ, ρ and r discussed in previous sections are square matrices. The depth takes discrete values, say $0, 1, 2, \ldots, n, \ldots$, and $\mathfrak{S}(n, m)$ is the scattering matrix associated with a layer extended from thickness n to m. The star-semi group properties hold as long as the principle of invariance is valid. The discrete case leads to a recursive relationship which does not apply in the continuous case.

For recursive relationships we consider $\widetilde{\mathfrak{S}} = \widehat{\mathfrak{S}}^{-1}$. For the discrete case,

$$\widetilde{\mathfrak{S}}(n) = \begin{pmatrix} t(n) - \rho(n)\tau^{-1}(n)r(n) & \rho(n)\tau^{-1}(n) \\ \tau^{-1}(n)r(n) & \tau^{-1}(n) \end{pmatrix} \tag{6.51}$$

$k, n = 0, 1, 2, \ldots$. After some algebra, we obtain

$$\widetilde{\mathfrak{S}}(0, n+1) = \widetilde{\mathfrak{S}}(n, n+1) \cdot \widetilde{\mathfrak{S}}(0, n) \tag{6.52}$$

and solving for $t(0, n+1)$, $\tau(0, n+1)$, $\rho(0, n+1)$, $r(0, n+1)$ in terms of $t(0, n)$, $\tau(0, n)$, $\rho(0, n)$, $r(0, n)$, we have a set of *recursive solutions*,

$$
\begin{aligned}
t(0, n+1) &= A(n)t(0, n) - [A(n)\rho(0, n) + B(n)][C(n)\rho(0, n) + D(n)]^{-1} \\
&\quad \times C(n)t(0, n), \\
\tau(0, n+1) &= \tau(0, n)[C(n)\rho(0, n) + D(n)]^{-1} \\
\rho(0, n+1) &= [A(n)\rho(0, n) + B(n)][C(n)\rho(0, n) + D(n)]^{-1}
\end{aligned}
\tag{6.53}
$$

and

$$r(0, n+1) = r(0, n) - \tau(0, n)[C(n)\rho(0, n) + D(n)]^{-1}C(n)t(0, n)$$

where

$$\begin{pmatrix} A(n) & B(n) \\ C(n) & D(n) \end{pmatrix} = \widetilde{\mathfrak{S}}(n, n+1) . \tag{6.54}$$

The relationships in eq. (6.53) have *initial value*

$$t(0, 0) = \tau(0, 0) = E \quad \text{and} \quad \rho(0, 0) = r(0, 0) = 0 . \tag{6.55}$$

This analysis provides us with a tool to compute the transmissions and reflections of arbitrary combinations of media from those of individual ones. This is called the *(right) sweep method or layer-peeling method* [12] *in computation.*

One may ask the question of existence of $[C(n)\rho(0, n) + D(n)]^{-1}$ in the above relationships. It can be shown that $[E - r(n, n+1)\rho(0, n)]$ has an inverse if and only if $[C(n)\rho(0, n) + D(n)]$ has an inverse. This means that in the discrete case the existence of the $*$ product implies the existence of recursive relations and vice versa.

Similar to eq. (6.51), for $\widehat{\mathfrak{S}}$ as defined in (6.40), we have

$$\widehat{\mathfrak{S}}(0, n+1) = \widehat{\mathfrak{S}}(0, n) \cdot \widehat{\mathfrak{S}}(n, n+1) . \tag{6.56}$$

As above, we lead to the so-called *(left) sweep method in computation.*

6.8 The Time-Dependent Case

The intensity $I^\pm = I^\pm(x,T)$ is a function of depth x and time T and likewise for the generator $M = M(x,T)$ in the time-dependent case. To specify the behavior of a thin medium, let the medium extend from x to $x+\Delta x$ where Δx is small. On physical grounds one would expect the transmission to involve two separate effects, as in the stationary case. First there should be a specular part of the incident intensity transmitted directly with a time delay λ which is Δx divided by the speed of light in the medium. This effect is described within $o(\Delta x)$ by

$$I^+(x + \Delta x, T_1) = I^+(x, T_1) - \lambda(x)\frac{\partial}{\partial T_1}I^+(x, T_1)\Delta x \ . \qquad (6.57)$$

There should also be a diffuse part, described by an equation of the form

$$I^+(x + \Delta x, T_1) = \left(\int^{19} b(x, T_1, T)I^+(x, T)\,dT \right)\Delta x + o(\Delta x) \qquad (6.58)$$

and thus specified by a kernel b. The total effect is given by superposition. Similar remarks apply to the reflection operator r, except that it is wholly specified by a diffuse term, associated with some kernel a. If (d, c, ν) are introduced for τ and ρ, analogous to the above (b, a, λ) for t and r, we see that the *generator* for the time-dependent case is

$$M = M(x, T_1) = \begin{pmatrix} b - \lambda\frac{\partial}{\partial T_1} & a \\ c & d - \nu\frac{\partial}{\partial T_1} \end{pmatrix} . \qquad (6.59)$$

By using the generator and taking the limit, we obtain

$$\frac{\partial}{\partial x}\mathit{\Pi}(x, T) = \begin{pmatrix} 1 & 0 \\ 0 & -1 \end{pmatrix} M(xT)\mathit{\Pi}(x, T) \qquad (6.60)$$

where $\mathit{\Pi}(x, T) = \begin{pmatrix} I^+(x, T) \\ I^-(x, T) \end{pmatrix}$.

Of special interest is Kaplan's equation for light scattering in a non-steady-state medium, see [8],

$$\frac{\partial I^+(x, T)}{\partial x} + \lambda\frac{\partial I^+(x, T)}{\partial T} = -l(x, T)I^+(x, T) + B(x, T) \qquad (6.61)$$

$$\frac{\partial I^-(x, T)}{\partial x} - \lambda\frac{\partial I^-(x, T)}{\partial T} = l(x, T)I^-(x, T) - B(x, T)$$

with

$$B(x, T) = \frac{1}{2x}\sigma(x, T)\int_{T_0}^{T} I^+(x, T') + I^-(x, T') \cdot \exp\left(-\frac{T - T'}{x} \right)dT' \ .$$

By comparing the above equation with eq. (6.60) and setting $\lambda = \nu$, we get

$$b(x, T_1, T_2) \quad = \quad d(x, T_1, T_2) = -l(x, T_1)\delta(T_1 - T_2) \qquad (6.62)$$
$$+\frac{\sigma(x, T_1)}{2x} \exp\left(-\frac{T_1 - T_2}{x}\right)$$

and

$$a(x, T_1, T_2) = c(x, T_1, T_2) = \frac{\sigma(x, T_1)}{2x} \exp\left(-\frac{T_1 - T_2}{x}\right)$$

for $T_1 \geq T_2$. The values are zero for $T_1 < T_2$. In eq. (6.62) $\sigma(x, T)$ is the scattering coefficient, l is the duration of temporal capture and $0 \leq T_0 \leq T' \leq T$.

It would be a mistake to think that the coefficients are always given by this direct method. Reference [11] and another part of this chapter give another derivation. The point to be emphasized is that the scattering matrix approach uses the generator to determine the mathematical structure.

The basic operations involving time-dependent operators are the sum and the product. The sum of two operators is defined in the usual way, by adding their values. But the product is defined by

$$(fg)(x_1, x_2, T_1, T_2) = \int f(x_1, x_2, T_1, T)g(x_1, x_2, T, T_2) \, dT . \qquad (6.63)$$

The subset of differentiable operators satisfies

$$\begin{array}{ll}
(fg)_{x_1} = fg_{x_1} + f_{x_1}g & (fg)_{x_2} = fg_{x_2} + f_{x_2}g \\
(fg)_1 = f_1 g & (fg)_2 = fg_2
\end{array} \qquad (6.64)$$

where subscripts x_1, x_2, 1 and 2 denote partial derivatives with respect to x_1, x_2, T_1 and T_2.

With these conventions, the *linear system* for intensities with general M, see eq. (6.59), is

$$\begin{pmatrix} I_h^+ + \gamma I_1^+ \\ I_h^- - \nu I_1^- \end{pmatrix} = \begin{pmatrix} b & a \\ -c & -d \end{pmatrix} \begin{pmatrix} I^+ \\ I^- \end{pmatrix} . \qquad (6.65)$$

In many applications, the coefficient $f(h_1, h_2, T_1, T_2)$ measures the contribution at time T_1 of a disturbance at time T_2. This is seen in such formulas as

$$I^+(x_2, T_2) = \int f(x_1, x_2, T_2, T_1)I^+(x_1, T_1) \, dT_1 . \qquad (6.66)$$

The f is called "*causal*" if the outputs at a certain time instant are dependent on the present and the past of the input only, i.e.,

$$f(x_1, x_2, T_1, T) = 0 \qquad \text{for } T_1 < T . \qquad (6.67)$$

We discovered that a weaker condition is sufficient to complete all analysis in the following sections.

Definition. The operator pair (f, g) is called *nonpredictive* [11] if

$$fg_1 + f_2g = 0 \; . \tag{6.68}$$

It should be observed that operators in stationary cases are always non-predictive since the time derivatives of g_1 and f_2 are zero. The nonpredictive condition is not needed for the derivation of eq. (6.65). It is essential in the following derivation.

Instead of intensities, let two nontrivial operators p and q satisfy eq. (6.65). Suppose that the integral equation $rp = q$ can be solved for r such that (r,p) is nonpredictive. Referring to eq. (6.64), we use

$$r_{x_1}p + rp_{x_1} = q_{h_1} \quad \text{and} \quad r_1p = q_1 \tag{6.69}$$

to compute

$$
\begin{aligned}
(-r_{x_1} + \nu\tau_1 - \lambda r_2)p &= rp_{x_1} - q_{x_1} + \gamma r_2 p \\
&= r(bp + aq - \lambda p_1) + (cp + dq) - \lambda r_2 p \quad (6.70) \\
&= (rb + rar + c + dr)p - \lambda(rp_1 + r_2 p) \\
&= (c + rb + br + rar)p \; .
\end{aligned}
$$

But $rp_1 + r_2p = 0$ because the operators are nonpredictive.

For the next equation, we assume that $tp = f$, where $f = f(x_1, x_2, T_1, T_2)$ is an operator independent of x_1, then

$$tp_{x_1} + t_{x_1}p = 0 \; .$$

As before, if we also assume (t,p) is nonpredictive, then

$$
\begin{aligned}
(-t_{x_1} - \lambda t_2)p &= tp_{x_1} - \gamma t_2 p \\
&= t(bp + aq - \mu p_1) - \gamma t_2 p \quad (6.71) \\
&= t(b + ar)p - \gamma(tp_1 + t_2 p) \\
&= t(b + ar)p \; .
\end{aligned}
$$

Two more equations for ρ and τ can be obtained if (t,p) and (r,p) are nonpredictive and if we use the expressions

$$\rho = -tp \quad \text{and} \quad \tau = q - rp \; . \tag{6.72}$$

Combining all four equations and cancelling non-singular p, we obtain the *time-dependent transport equations*

$$-\frac{\partial}{\partial x_1}\mathfrak{S} = C + D\mathfrak{S} + \mathfrak{S}B + \mathfrak{S}A\mathfrak{S} \tag{6.73}$$

where $C = M_{12}$, $D = M_{11}$, $D = M_{22}$ and $A = M_{21}$ with M_{ij} obtained from M, eq. (6.59), by replacing elements in the i-th row and the j-th column by zero operators. For example, the *left-hand reflection operator* $r = r(h_1, h_2, T_1, T_2)$ satisfies an integro-differential equation of the form

$$-r_{x_1} + \nu\tau_1 - \lambda\tau_2 = c(x_1, T_1, T_2) + \int d(x_1, T_1, T)r(x_1, x_2, T, T_2)\, dT$$

$$+ \int r(x_1, x_2, T_1, T)b(x_1, T, T_2)\, dT \qquad (6.74)$$

$$+ \int\int r(x_1, x_2, T_1, T)a(x_1, T, T')$$

$$\times r(x_1, x_2, dTdT', T', T_2) .$$

For the plane parallel layer case

$$
\begin{array}{ll}
a = \frac{1}{4\pi\mu}\gamma^{-+}, & b = \frac{1}{\mu}\delta e^{-\alpha h/\mu} + \frac{1}{4\pi\mu}\gamma^{++} \\
c = \frac{1}{4\pi\mu}\gamma^{+-}, & c = \frac{1}{\mu}\delta e^{-\alpha h/\mu} + \frac{1}{4\pi\mu}\gamma^{--} ,
\end{array}
\qquad (6.75)
$$

where δ, α and $\gamma^{\pm\pm}$ are respectively the Dirac delta, the time-dependent attenuation coefficient and the normalized phase operators.

For the *time-invariant case*, the operators t, τ, ρ, r depend only on the difference in time $T_2 - T_1$. Then the left-hand reflection operator

$$r = r(x_1, x_2, T_2 - T_1) \qquad (6.76)$$

and

$$rI^+(x_2, T_2) = \int r(x_1, x_2, T_2, T_1)I^+(x_1, T_1)\, dT_1 \qquad (6.77)$$

$$= \int r(x_1, x_2, T_2 - T_1)I^+(x_1 T_1)\, dT_1 .$$

Eq. (6.73) is reduced to the stationary case by taking the Laplace transform $\bar{r}$ of r subject to a mild smoothness condition, $r(h_1, h_2, T)e^{sT} \to 0$ as $T \to \infty$. Then $\bar{r}$ satisfies the equation

$$
\begin{aligned}
-\bar{r}_{x_1} =\ & \bar{c}(x_1, T_2 - T_1) + [\bar{d}(x_1, T_2 - T) + s(\vartheta + \lambda)]\bar{r}(x_1, x_2, T - T_2) \\
& + \bar{r}(x_1, x_2, T_1 - T)\bar{b}(x_1, T_1 - T_2) \qquad (6.78) \\
& + \bar{r}(x_1, x_2, T_1 - T)a(x_1, T - T')\bar{r}(x_1, x_2, T' - T_2) .
\end{aligned}
$$

Likewise, the Laplace transform $\bar{t}$ of t satisfies the equation

$$
\begin{aligned}
-\bar{t}_{x_1} =\ & -s\lambda\bar{t}(x_1, x_2, T_2 - T_1) + \bar{t}(x_1, x_2, T_1 - T')\bar{b}(x_1, T' - T_2) (6.79) \\
& + \bar{t}(x_1, x_2, T - T')\bar{a}(x_1, T - T')\bar{r}(x_1, x_2, T' - T_2) .
\end{aligned}
$$

Eq. (6.74) is a generalized Riccati equation. It is well-known that the *cross ratio* is a constant in the scalar Riccati equation. This property is useful to establish relations among four distinct solutions, i.e., one solution is determined by any three other solutions.

Let u and v satisfy the respective equations

$$-u_{x_1} + \nu u_1 = \bar{A}u \qquad v_{x_1} - \lambda v_2 = v\overline{B} \qquad (6.80)$$

for some operators $\bar{A}$ and $\bar{B}$. Then it can be easily shown that

$$\omega = ugv \tag{6.81}$$

satisfies the equation

$$-\omega_{x_1} + \nu\omega_1 - \lambda\omega_2 = \bar{A}\omega + \omega\bar{B} \tag{6.82}$$

if and only if

$$ug_{x_1}v = 0 . \tag{6.83}$$

Suppose that r^i and r^j both satisfy eq. (6.74). By subtraction and rearrangement $\omega = r^i - r^j$ satisfies eq. (6.82), with

$$\bar{A} = d + r^j a \qquad \bar{B} = b + ar^i . \tag{6.84}$$

Let $u = \tau^i$ denote the value of τ belonging to the system of eq. (6.73) when $r = r^i$ and $v = t^i$. Then eq. (6.80) holds with values of $\bar{A}$ and $\bar{B}$ in eq. (6.84), and hence

$$r^i - r^j = \tau^i g^{ij} \tau^i . \tag{6.85}$$

Assume next that r^1, r^2, r^3, r^4 are four solutions of eq. (6.74) which are distinct in the sense that $r^i - r^j$ is nonsingular for $i \neq j$. Then setting $\tau = \tau^1$, we have the cross ratio,

$$X = (r^4 - r^3)(r^4 - r^2)^{-1}(r^3 - r^2)(r^3 - r^1)^{-1} = \tau g \tau^{-1} \tag{6.86}$$

where g is independent of x_1.

If τ^1 and g commute, we get $X = g$, a generalization of a well-known property of the scalar Riccati equation. It follows from the properties of the cross-ratio that if r^1, r^2 and r^3 are any three particular solutions of eq. (6.74), then the *general solution* is expressed in the form

$$r = (E - K)^{-1}(r^3 - Kr^2) \tag{6.87}$$

with $K = \tau g \tau^{-1}(r^3 - r^1)(r^3 - r^2)^{-1}$, where $K \neq E$.

Let us consider a time invariant plane parallel layer divided into k sub-media, the i-th sub-medium with h extending from x_{n-1} to x_n, $n = 1, 2, \ldots, k$. Points $x_0, x_1, \ldots, x_k$ are chosen so that in each sub-medium the one way travel time is one unit in both the forward and backward directions. Besides the assumption that $\lambda = \nu$ and time invariance, this is a general model. With respect to the two inputs $I^+(n)$ and $I^-(n+1)$, the two outputs are delayed by one unit of time. The outputs are

$$\begin{pmatrix} I^+(n+1, T+1) \\ I^-(n, T+1) \end{pmatrix} = \begin{pmatrix} \delta + b\Delta & a \\ c & \delta + d\Delta \end{pmatrix} \begin{pmatrix} I^+(n, T) \\ I^-(n+1, T) \end{pmatrix}$$

$$\text{output} \qquad\qquad\qquad\qquad\qquad\qquad \text{input}$$

$$\tag{6.88}$$

This mathematical expression can be illustrated by figure 6.5.

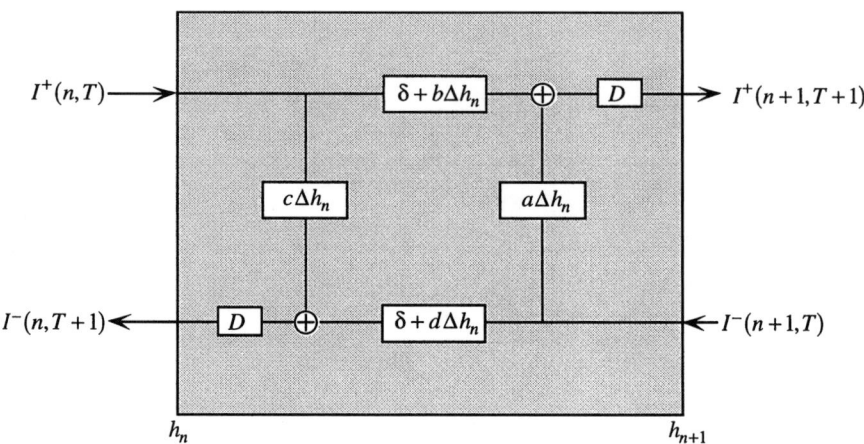

Figure 6.5. Time-dependent inputs and outputs of a thin medium

The figure 6.5 indicates that if there is only an input $I^+(n,T)$ at x_n, as it enters the medium part of this input is transmitted in the amount $(\delta + b\Delta x_n)I^+(n,T)$ and followed by a delay operator D. Also part is reflected by the amount $(c\Delta x_n)I^+(n,T)$ and a delay D, defined as

$$Df(T) = f(T+1) \ . \tag{6.89}$$

Likewise for the input $I^+(n+1,T)$ at x_{n+1}. Linear combination is used when there are two inputs.

In radiative transfer in plane parallel atmosphere volume attenuation and the normalized phase function are independent of time T. Then the 2×2 square matrix in eq. (6.88) is the scattering matrix considered in the stationary case. So for the *discrete time-dependent case* the *scattering matrix*

$$\mathfrak{S}(n, n+1) = \begin{pmatrix} D & 0 \\ 0 & D \end{pmatrix} \begin{pmatrix} \delta + b & a \\ c & \delta + d \end{pmatrix} \Delta x_n \ . \tag{6.90}$$

The nonpredictive operator introduced in eq. (6.68) for (r, ρ) now can be expressed as

$$r(T)\rho(T+1) = r(T+1)\rho(T) \ . \tag{6.91}$$

By using the delay operator and the associative law,

$$[r(T)D]\rho(T) = [Dr(T)]\rho(T) \tag{6.92}$$

for every $\rho(T)$, then we have the *commutative law*

$$rD = Dr \ . \tag{6.93}$$

Likewise the commutative law is true for t, τ and ρ, as long as they are non-predictive and time-invariant. The commutative law also holds for D^{-1}, which is defined as

$$D^{-1}f(T) = f(T-1) .$$ (6.94)

Eq. (6.88) also can be obtained by replacing partial derivatives by finite differencing. With the above analysis and referring to (6.75), eq. (6.88) for discrete time-dependent radiative transfer has the form

$$\begin{pmatrix} I^+(n+1, T+1) \\ I^-(n, T+1) \end{pmatrix} = \left[\begin{pmatrix} \frac{1}{\mu}\delta & 0 \\ 0 & \delta\frac{1}{\mu} \end{pmatrix} e^{-(\alpha/\mu)\Delta x_n} \right.$$
$$\left. + \frac{1}{4\pi\mu} \begin{pmatrix} \gamma^{++} & \gamma^{-+} \\ -\gamma^{+-} & -\gamma^{--} \end{pmatrix} \right] \begin{pmatrix} I^+(n, T) \\ I^-(n+1, T) \end{pmatrix} .$$ (6.95)

As in the stationary case, the discrete time-dependent $\tilde{\mathfrak{S}}$, see eq. (6.51), can be constructed by using $\mathfrak{S}$. The result is

$$\tilde{\mathfrak{S}} = \tilde{\mathfrak{S}}(n, n+1) = \begin{pmatrix} E & 0 \\ 0 & D^{-1} \end{pmatrix} \begin{pmatrix} \delta + b\Delta x_n & a\Delta x_n \\ -c\Delta x_n & \delta - d\Delta x_n \end{pmatrix} \begin{pmatrix} D & 0 \\ 0 & E \end{pmatrix} .$$ (6.96)

As we see, $\tilde{\mathfrak{S}}$ does not involve the inverse operator with the exception of the operator D^{-1}.

With the associative property and D & D^{-1} commuting with all operators in $\tilde{\mathfrak{S}}$, it is easy to see that eq. (6.52) holds for $\tilde{\mathfrak{S}}$, and one can apply the right sweep method. In a similar way, the discrete time dependent, see eq. (6.40),

$$\hat{\mathfrak{S}} = \hat{\mathfrak{S}}(n, n+1) = \begin{pmatrix} D^{-1} & 0 \\ 0 & E \end{pmatrix} \begin{pmatrix} \delta - b\Delta x_n & a\Delta x_n \\ c\Delta x_n & (\delta + d)\Delta x_n \end{pmatrix} \begin{pmatrix} E & 0 \\ 0 & D \end{pmatrix}$$ (6.97)

which is related to the right sweep method, by eq. (6.54) and (6.55), in computation. Similarly, we have the left sweep method for the time-dependent case.

6.9 Concluding Remarks

In this chapter we introduced the scattering matrix which governed the inputs and outputs in a plane parallel layer. The concept is basic and many new mathematical results are presented here. Results can be applied to many physical systems including radiative transfer.

After some elementary discussion of algebra of scattering matrices, we constructed the transport equations, involving operators and generators (or coefficients), from the invariant imbedding technique. We give the n term solutions for such equations in the form of Bremmer series. For computation, we discussed the discrete scattering matrices which leads to sweep methods and others used in this book. The scattering matrices and transport equations can be easily extended to the time-dependent, continuous and discrete cases.

References

1.   R. Bellman, H. Kagiwada, R. Kalaba, and S. Ueno, "Invariant Imbedding and the Computation of Internal Field for Transport Processes," *J. Math. Anal. Appl.*, Vol. 12, 1965, pp. 541–548.

2.   R. Bellman, H. Kagiwada, R. Kalaba, and S. Ueno, "The Invariant Imbedding Equation for the Dissipative Function of an Inhomogeneous Finite Slab with Anisotropic Scattering," *J. Math. Physics*, Vol. 8, 1967, pp. 2137–2142.

3.   R. Redheffer, "On the Relation of Transmission Line Theory to Scattering and Transfer," *J. Mathematics and Physics*, Vol. XLI, No. 1, March 1962, pp. 1–41.

4.   S. Chandrasekhar, *Radiative Transfer*, Dover Publications, Inc., New York, 1960.

5.   A. P. Wang, "The Four-Port Dynamic System," *J. Math. Anal. & Appl.*, Vol. 120, No. 2, 1986, pp. 416–437.

6.   A. P. Wang, "Conjugacy and Boundary of an Extended Scattering System," *J. Math. Anal. & Appl.*, Vol. 133, No. 2, August 1, 1988, pp. 383–394.

7.   A. P. Wang, "Dissipative Properties of a Nonlinear Operator Equation," *J. Math. Anal. & Appl.*, Vol. 72, No. 1, Nov. 1979, pp. 75–88.

8.   S. A. Kaplan, "On the Theory of Light Scattering in a Non-steady-state Medium," *Astron. J. U.S.S.R.*, Vol. 39, 1962, pp. 702–709.

9.   D. Dudley and A. P. Wang, "Operator Theory on WKB Method and Bremmer Series," *J. Math. Physics*, Vol. 24, No. 6, June 1983, pp. 1470–1476.

10.   R. Bellman, R. Kalaba and Sueo Ueno, "Invariant Imbedding and Time-dependent Diffused Reflection of a Pencil of Radiation by a Finite Inhomogeneous Flat Layer, Part I," *J. Math. Anal. Appl.*, Vol. 7, 1963, pp. 310–321.

11.   R. M. Redheffer and A. P. Wang, "Formal Properties of Time-Dependent Scattering Processes," *J. Math. and Mechanics*, Vol. 19, No. 9, 1970, pp. 765–781.

12.   A. M. Bruckstein, B. C. Levy and T. Kailath, "Differential Methods in Inverse Scattering," *SIAM J. Appl. Math.*, Vol. 45, No. 2, April 1985, pp. 312–335.

7. Atmospheric Correction

The apparent radiance of ground reflection as measured by a remote sensor differs from the intrinsic surface radiation because of the presence of the intervening atmosphere. Methods are developed in this chapter to remove the atmospheric effects from measured data taken in air and in space.

7.1 Introduction

In recent years, with the advent of advanced earth monitoring space- and air-craft, it has become increasingly important to evaluate the extent of atmospheric effects on remotely sensed data, because the terrestrial atmosphere tends to diminish the ability to discriminate between target and surroundings. In other words, apparent ground radiance distribution obtained by space- or air-borne sensors differs from the intrinsic ground radiance distribution, because of the multiple scattering of radiation in the atmosphere-ground system. The atmosphere can selectively scatter, absorb, re-emit, and refract radiation that traverses through it. In this context, the atmosphere has a filtering or distorting function that changes spatially, spectrally, and temporally.

Hence, an allowance for the atmospheric effects should improve the accuracy of pattern recognition and image interpretation in global monitoring. To insure the usefulness of remote sensing, it is necessary to determine the atmospheric effects due to scattering, absorption and reflection on the radiation field of the atmosphere-ground system. Evaluations of the atmospheric effect based on the horizontally uniform radiation field have been attempted by several authors (cf. [1–8]). In the case of a horizontally non-uniform atmosphere, the scattering effect has been analytically studied (cf. [9–11]).

In the real atmosphere-ground system, the assumption of the horizontally uniform radiation field is seldom realized. Even if the atmosphere were spatially uniform over an extended area, non-uniform atmospheric effects must be present because of the non-uniform ground albedo (what we may call the 'blurring effect'). Several attempts have been made to take into account the diffuse radiation field due to the horizontally non-uniform ground albedo, with the aid of the adding-and-doubling procedure, the Fourier transform method, the invariant imbedding technique and others (cf. [12–21]).

In the present chapter, based on the three-dimensional transfer model consisting of a one-dimensional free atmosphere bounded by a horizontally non-uniform diffuse reflector, we express the required total spectral radiance at the top in terms of the scattering and transmission functions for the free atmosphere, and the horizontally inhomogeneous ground albedos.

In the second section it is shown that the total spectral radiance recorded by the remote sensor can be expressed in terms of the free-atmospheric path radiance, the directly and diffusely transmitted components of the surface radiance. Among them the diffusely transmitted component of the surface radiance gives rise to the blurring of the ground image. Two approaches for correcting the atmospheric effects are outlined. It should be noted that the correction procedure for aircraft data is much more complicated in comparison with that for satellite data, because the computation of the internal radiation field requires a large amount of numerical work and in addition we have to evaluate the scan angle effect properly. Furthermore, realistic optical models of the atmosphere are established, allowing for the scattering and absorption due to the molecular gas and the stratospheric and coastal types of aerosols. In the third section, the procedure of correcting the blurring effect on satellite data is described, and the procedure is applied to a LANDSAT data set, for a series of atmospheric models with different haze levels. In the fourth subsection, we describe a somewhat simplified correction procedure that is appropriate for aircraft data. This procedure is applied, using the standard atmosphere, to a data set obtained by Japan Research Committee of Environment Remote Sensing (JRCERS).

Finally, a general solution and approximations are presented in the last section utilizing the theory of Chapter 6. This general solution considers all multi-scattering between ground and atmosphere; the view angle is not limited to the normal to the ground.

7.2 Radiative Processes in the Atmosphere

7.2.1 Diffuse radiance in the atmosphere

The radiative processes in the atmosphere bounded by a non-uniform reflector may be described schematically by Fig. 7.1. The diagram represents the case of satellite observation. The detector above the atmosphere measures the upward radiance in the direction of its line of sight. Since the range of the view angle by a space-borne sensor is small, we may consider only the upward normal radiance. In the case of aircraft observation there are two significant differences: 1) the detector is within the atmosphere, 2) the scan angle effect must be considered because of the large range of the view angle. When the space-borne or air-borne sensor sees a target on the background, the following three types of photons reach the detector:

Type A   Photons reflected by the target and then directly transmitted by the free atmosphere.

Type B   Photons transmitted diffusely by the free atmosphere, after having interacted with the background.

Type C   Photons reflected diffusely by the free atmosphere.

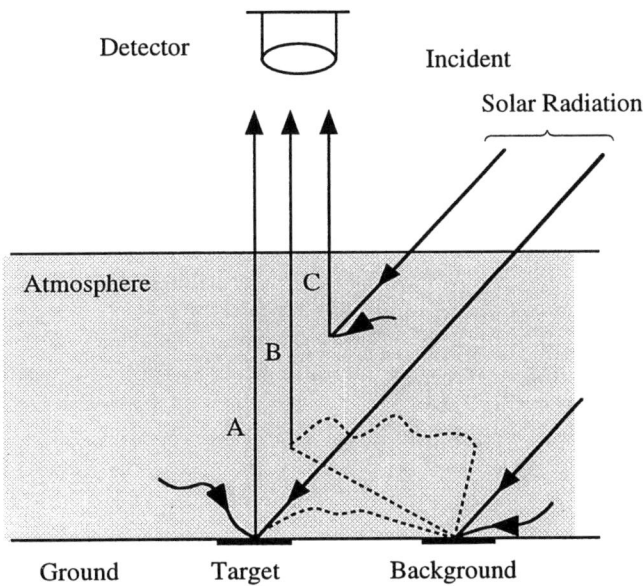

Figure 7.1. The radiative processes in the atmosphere with a non-uniform reflector.

It may be said that Type-A photons carry the direct information on the target, Type-B photons represent the combined effects by both the background and atmosphere, and Type-C photons represent the free-atmospheric effects. The radiance due to Type-B and Type-C photons is usually called the path radiance. In particular, Type-B photons produce the blurring of the ground patterns in much the same way as instrumental blurring. The radiative processes those photons undergo before reaching the detector are as follows: 1) direct transmission through the atmosphere (shown by solid straight lines in Figure 7.1), pure atmospheric diffusion (solid wavy lines), and 2) atmospheric diffusion involving ground reflection (dotted wavy lines).

A dotted straight line is used to represent direct transmission after ground reflection.

The presence of the path radiance obscures the intrinsic surface image. Therefore, it is very important to remove these types of radiance and deal only with the radiance by Type-A photons. In order to eliminate the contributions by both Type-B and Type-C photons exactly from the observed radiance, we must solve a three-dimensional radiative transfer problem in the atmosphere-ground system, and it would be an extremely difficult task.

Instead of solving the exact problem, we formulate approximate atmospheric correction methods which are applicable to the real three-dimensional environment. These approximation methods utilize the results from one-dimensional radiative transfer models.

In the first correction method (the convolution method) we obtain a convolution-type integral equation relating the albedo distribution to the observed radiance map, while the kernel of the equation is given by the properties of the atmosphere. Although this method is considered to be more exact than the second one, the numerical procedures for applying it to aircraft data would be much more complicated. The second correction method (the average method) is found to be applicable to the multi-spectral scanner (MSS) data obtained by the space-borne sensor, as well as to those obtained by the air-borne sensor. This method was used to remove the atmospheric effect in our previous work [14]. To complete the analysis, we construct general models which consider all multi-scattering (the scattering matrix method) and we obtain an asymptotic solution.

7.2.2 Atmospheric models

In order to solve the radiative transfer equation, we must have necessary atmospheric parameters. In this investigation we shall use the atmosphere of Elterman [22] as a standard model of the earth's atmosphere. This is a reasonable choice, at least for our test imagery, because it is apparently cloud free. Since it is known that the characteristics and the size distribution of the aerosols are different in the lower atmosphere (altitude < 15 km) from those in the upper atmosphere (> 15 km), we divide the atmosphere into two layers. The ground is assumed to be a Lambertian surface. The necessary atmospheric parameters, such as the optical thickness of the molecular gas, that of the aerosols, the turbidity factor of the molecular gas, and the absorption by ozone in each layer of the atmosphere, are provided in Elterman's table.

We use the Haze M and Haze H for the types of the aerosols in the lower and the upper layers of the atmosphere, respectively. The Haze M represents maritime- and coastal-type aerosols. Since our study site is in the coastal area, it is justified to use the Haze M in the lower layer. The Haze H is the stratospheric type of aerosols. Discussions on the aerosol types, including Haze M and Haze H, are given by Deirmendjian [23]. The single scattering phase functions for the Hazes M and H show a strong forward scattering pattern and may be computed by means of Mie's theory. The single scattering phase function for a molecular particle is given by Rayleigh's scattering formula. In solving the one-dimensional radiative transfer problem, we treat the multiple scattering exactly for a three-layer model by using the (one-dimensional) doubling and adding methods. These methods have been commonly used for radiative transfer in planetary atmospheres by Hansen and Travis, [24].

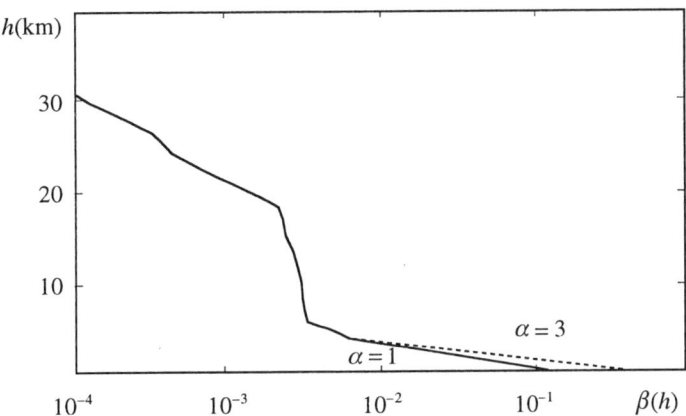

Figure 7.2. The aerosol attenuation profiles.

The solid curve in Figure 7.2 shows the attenuation profile due to aerosol particles, at the wave-length of 0.55μm, as taken from Elterman's table. Corresponding profiles for other wavelengths have similar characteristics. It should be noted that the attenuation coefficient decreases exponentially as the altitude increases in the first 4 kilometers. It will be a reasonable assumption that the haze level (or visibility) of the atmosphere is in most cases determined by the aerosols, which are concentrated heavily in the lowermost portion of the atmosphere. It is also noted that the concentration of the aerosols dominates that of the molecules in this portion of the atmosphere. Therefore, in order to account for the variation of the haze level atmosphere, we extend the standard aerosol profile by changing the gradient of the straight-line portion (in the semi-logarithmic diagram), in a manner shown by the dotted lines. For 0.55μm, this portion of the aerosol attenuation profile, $\beta(h)$, is numerically expressed as

$$\beta(h) = 0.158e^{-0.7916h}\alpha^{1-h/4}. \tag{7.1}$$

Here, h is in units of kilometer. Similar expressions (having different multiplying factors) have been obtained for other wavelengths. In the standard case of Elterman, we have $\alpha = 1$.

7.3 Atmospheric Correction for Landsat Data

7.3.1 Single-reflection approximation

We assume that, for a given wavelength, the atmosphere is horizontally uniform. Let us denote by $\pi F f_G$ the downwards flux at the bottom of the free atmosphere, i.e., the amount of radiation reaching the ground, either directly or diffusely transmitted through the atmosphere, where the input solar radiation is given by πF. When we are dealing with LANDSAT data [16], it is only necessary to derive the expression for the vertical component of the radiation emerging from the top of the atmosphere, since in this case the scan angle of observation is sufficiently small.

In the single reflection approximation, the second- (and higher-) order processes for those photons, which have had interactions with the ground, are neglected. The accuracy of this approximation has been checked, using a standard atmosphere with a uniform ground, see [16]. For the ground albedo of less than 0.5 (true for the usual ground materials), the errors are found to be several percent, small enough for most applications.

Under the geometry shown in Fig. 7.3, assuming the single reflection approximation, the total radiance detected by the space-borne sensor, $I_0(x, y)$, is expressed in terms of the quantities in observed units such that $I'_0 = C I_0$ and $I'_A = C I_A$, with the gain factor C; (see [19]).

$$I'_0(x, y) = I'_A(x, y) + \int_{-\infty}^{\infty} k(\xi, \eta) I'_A(x + \xi, y + \eta) d\xi d\eta. \qquad (7.2)$$

In Equation (7.2)

$$I'_A(x, y) = C F f_G A(x, y) e^{-\tau_0} \qquad (7.3)$$

where τ_0 is the optical thickness of the atmosphere and $A(x, y)$ is the ground albedo at the point (x, y). The function, $k(\xi, \eta)$, is defined by

$$k(\xi, \eta) = \frac{1}{4\pi} \int_0^H \frac{h}{(\xi^2 + \eta^2 + h^2)^{3/2}} \cdot P(\theta) e^{(1-1/\mu)\tau(h)} dh, \quad 0 \le h \le H, \quad (7.4)$$

Here $P(\theta)$ is the single scattering phase function, μ the direction cosine $\cos \theta$ and h the geometrical thickness of the atmosphere. It should be mentioned that Equation (7.2) is found, provided that our concern is only in the removal of the blurring (namely, the Type-B component), thus allowing for an undetermined constant in the solution.

In Equation (7.4), the kernel $k(\xi, \eta)$ can be computed, when the optical properties of the atmosphere are given. Then the distribution $I'_A(x, y)$ can

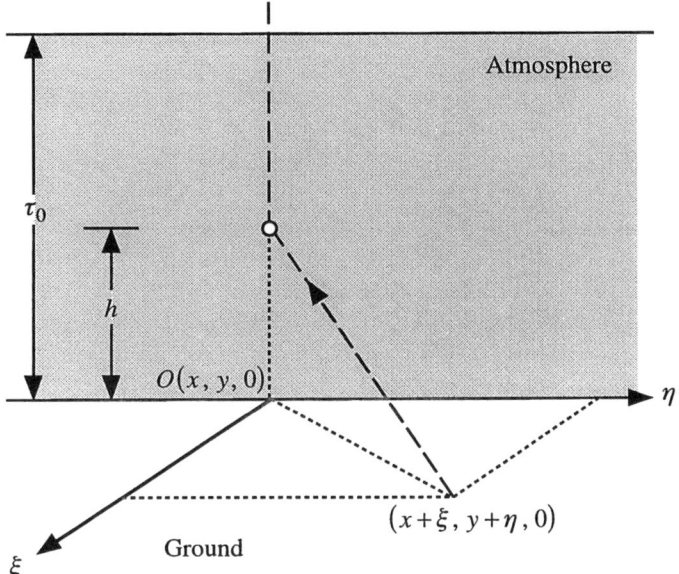

Figure 7.3. The diagram of the signal reflection and re-scattering approximation.

be determined for the observed radiance map, $I_0'(x, y)$. This latter quantity gives the blurring-free radiance map, aside from an undetermined additive constant.

7.3.2 Correction procedure (convolution method)

In order to solve Equation (7.2) for $I_A'(x, y)$, we consider the two-dimensional Fourier transform of the equation in discrete form:

$$O_{mn} = A_{mn} + K_{mn} A_{mn},\qquad(7.5)$$

where A_{mn}, O_{mn} and K_{mn} are the Fourier coefficients of order (m, n), of the quantities, $I_A'(x, y)$, $I_0'(x, y)$ and $k(\xi, \eta)$. From Equation (2.7), we obtain

$$A_{mn} = \frac{O_{mn}}{1 + K_{mn}}.\qquad(7.6)$$

This discrete form of $I_A'(x, y)$ is then given as

$$a_{kl} = \frac{1}{N^2} \sum_{m=0}^{N-1}\sum_{n=0}^{N-1} \left(\frac{O_{mn}}{1 + K_{mn}} \right) e^{2\pi j / N (km + ln)}.\qquad(7.7)$$

Here we have assumed that the observed image data are given as a square matrix of the $N \times N$ pixels. The indices (k, l) give the position of a pixel in the matrix.

7.3.3 Results and discussion

The convolution method described above has been applied to a LANDSAT data set, see [16], covering the central part of Honshu (the main island) of Japan. The portion of the data set we have studied consists of 256×256 pixels, including the City of Kanazawa and its vicinity (roughly corresponding to a 20 km × 20 km ground area). The geography consists of the sea, the river, the water channel, the wooded mountain area and the flat land. A sequence of atmospheric models, with $\alpha = 1$ (standard model), 3, and 10, is used in the correction procedure. The case with $\alpha = 10$ is unlikely for this particular observation (according to the weather forecast, the sky was 'clear' on that day). However, the case was included to examine the effect of over-correction.

In the case of Band 4, (corrected with $\alpha = 3$) we observe the following improvements: 1) The water channel is more clearly seen in the corrected image. 2) The river is better recognized in the corrected image. 3) There is a slight improvement in the ground patterns of the flat region. 4) There does not seem to be a significant difference in the wooded region.

7.4 Atmospheric Correction for Aircraft Data

7.4.1 Evaluation of the internal radiation field

In order to evaluate the atmospheric effects for the MSS data obtained by the air-borne sensor by JRCERS, it is necessary to consider an internal radiation field. We compute the necessary quantities of internal radiance at three different altitudes of 1km, 3km, and 50km. In the computation the atmospheric model of Elterman ($\alpha = 1$) is assumed to represent the Earth's atmosphere. We make computations for three different wavelengths ($\lambda = 0.475\mu$m, 0.575μm, and 0.670μm). They correspond to the central wavelengths of the particular three bands of aircraft MSS data supplied by JRCERS.

Let us introduce the following quantities for the evaluation of the scattering effects within the atmosphere. The direct upward radiance at a certain altitude due to Type-A photons is denoted by U_A, the corresponding other kinds of radiances due to Type-B and -C are denoted by U_B and U_C, respectively. The total upward radiance is given by the sum of the above three radiances, U_A, U_B and U_C. It is denoted by U_T. The sum of the radiances U_B and U_C is called the upward path radiance, denoted by U_P.

In Figure 7.4 we present the theoretically computed U_T as a function of the ground albedo and the scan angle. The linear dependence of U_T on the ground albedo A holds up to $A < 0.5$. The slope of the total upward radiance curve is flattened as the scan angle increases. The ratios U_A/U_T and U_P/U_T are shown as a function of the ground albedo A and the scan angle θ in Figure 7.5. The shapes of the right side portion and the left side portion of the curves are not symmetric about the line of zero scan angle. This indicates that the degree of the atmospheric effects depends on which side the sensor scans with respect to the forward direction of the moving aircraft. In the course of scanning from the right to the left, the transition of the azimuthal angle occurs at zero scan angle. In Figure 7.5 we plot the case of the azimuth transition from 50° to 130°. The asymmetric nature of the ratios U_A/U_T and U_P/U_T about zero scan angle is shown in Figure 7.5. It is due to atmospheric scatterings. It is, therefore, understood that the above asymmetry about zero scan angle becomes less with the increase of the ground albedo, or with the decrease of the altitude. The above asymmetry about zero scan angle may be simply explained by differences of the azimuthal angle.

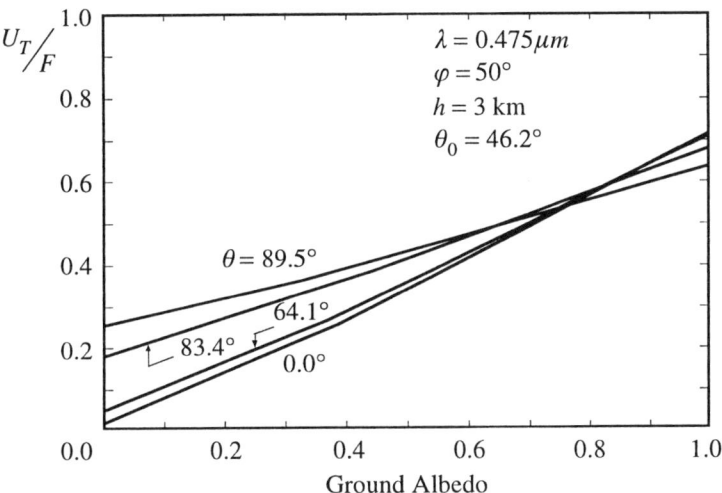

Figure 7.4. The total upward radiance as a function of the ground albedo and the scan angle.

7.4.2 Results and Discussion

In the application of the averaging method to real image data, we must estimate the size of the area over which the averaging should be performed. The radius of the area is found to be about 200km, corresponding to 25 pixels in the test image. The actual application is made to the imagery of the Kanazawa port area which was taken by JRCERS aircraft on Oct. 14, 1977. This area is located in the left central portion of the LANDSAT imagery used in Section 3. The studied imagery consists of 768 × 768 pixels, roughly corresponding to a 6km × 6km ground area. We observe the following differences when two map images before and after the atmospheric correction are compared. In the scattering-corrected image, thin surface structures appear more clearly. This is particularly true for a breakwater in the sea and a

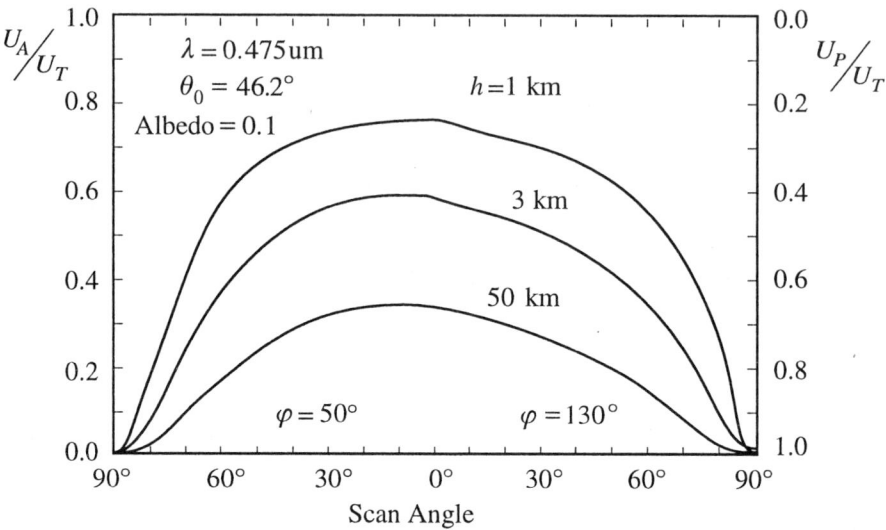

Figure 7.5. The internal radiance ratios as a function of the ground albedo and the scan angle.

coastal shore line. The border between the land and the water is more clearly recognized in the corrected image. Some of the roads may be found in the corrected image, while they are hardly recognized in the uncorrected image. The results are for the wavelength $\lambda = 0.575\mu$m; see [16] for more details.

It has been shown that the correction of the atmospheric effects for aircraft data can be accomplished by the removal of both the blurring effect and the scan angle effect. The application of the averaging method to real aircraft image data is made for the first time, and we obtain good results.

We conclude that the atmospheric correction of remotely sensed earth's imagery enables us to make a more accurate interpretation of the data than in the case where no such correction is made.

7.5 Outline of the AECS Software

The AECS is a system of computer software programs and performs corrections for the effects of the pure atmospheric path radiance, the blurring due to the interaction of photons with surface background materials and the view angle. The block diagram of the AECS is shown in Figure 7.6. The AECS consists of two subsystems named the Correction Subsystem and the Inter-Extrapolation Subsystem. In the Inter-Extrapolation Subsystem we have three blocks, namely, the Radiance File, the Inter-Extrapolation Program and the Correction Table. In the Radiance File we store the necessary information on the theoretical radiance quantities for typical atmospheric and observational conditions. To make this file we computed the theoretical values on the radiances for combinations of the following different cases. In the wavelength region we considered four wavelengths, i.e., $0.47\mu m$, $0.575\mu m$, $0.6\mu m$ and $0.845\mu m$, which correspond to observed wavelength used by Japan Research Committee of Environmental Remote Sensing's airborne sensor. As for the altitude, three cases of 1km, 3km, and 50km were considered. The first two altitude cases correspond to the typical airborne sensor's altitudes, whereas the case of 50km is applicable to LANDSAT data. We considered the cases of four atmospheric models with different optical thickness, corresponding to the cases from clear standard atmosphere to thick hazy sky. We also considered eleven different values of ground albedo, starting from 0.0 to 1.0 with an increment of 0.1. Thirteen different values of the incident sun angle and view angle were used in the computation for each combination of the above cases. A considerable amount of computer time is required for such computations. Once computations are carried out and the parameterized radiance coefficients are stored in the Radiance File, the Inter-Extrapolation Program computes the fraction of the path radiance to the total radiance as a function of the view angle quite efficiently.

Since it is found that most photons received by the sensor have at most two interactions with ground materials, both the theoretical total radiance I_T and the direct radiance I_D may be expressed in the quadratic function with respect to the ground albedo A.

$$I_T = P + QA + RA^2, \qquad (7.8)$$

$$I_D = SA + TA^2. \qquad (7.9)$$

These coefficients P, Q, R for the total radiance and S, T for the direct radiance are found by applying the least squares method to the computed radiance results. These parameterized coefficients for the radiances are stored in the Radiance File.

If input parameters, such as the optical thickness of the atmosphere, the flight altitude, wavelength and incident sun angle deduced from the longitude and the latitude of the observed site and the observed time are given, the fraction of the path radiance to the total radiance is computed by the Inter-Extrapolation Program and the results are stored in the Correction Table. Then the Correction Program accepts the unit scan angle, the flight altitude and the wavelength as inputs and uses the values stored in both the Correction Table and the Original Image File to produce the atmospheric effects free image. The Correction Program adopts the average method for the blurring and the pure atmospheric path radiance effects, and then performs the multiplication of inverse attenuation factor for the scan angle effect. The

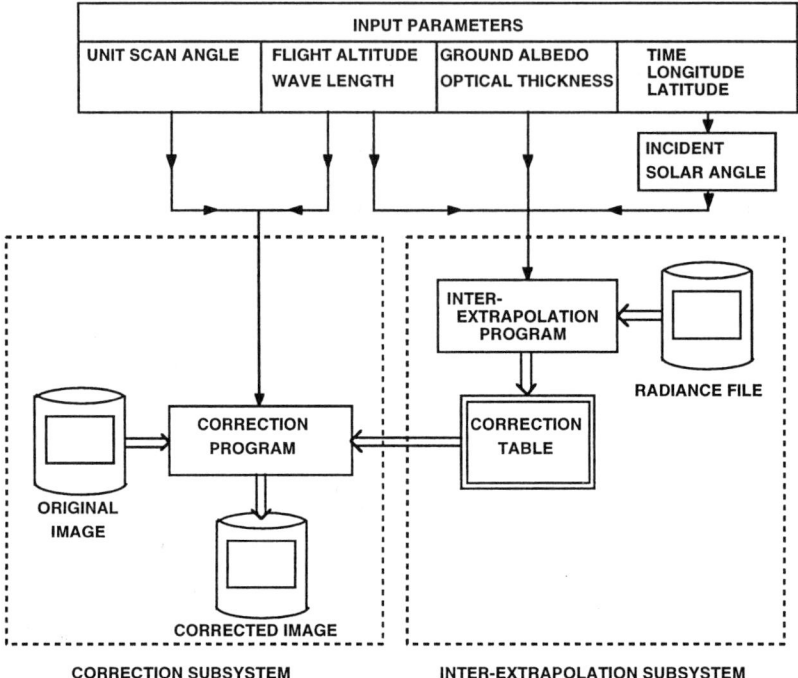

Figure 7.6. The block diagram of the AECS system.

information on both unit scan angle and the flight altitude is needed to establish the correspondence between the pixels in the image and the actual ground location.

With regards to the atmospheric corrections, one of the most important parameters is the atmospheric condition at the time of the observation. The atmospheric condition may be represented in terms of the atmospheric optical thickness. The block diagram for the estimation of the optical thickness is shown in Figure 7.7. The optical thickness should be given either by direct measurement or by the iterative estimation from observed data sets themselves. We may obtain the optical thickness of the atmosphere by measuring the horizontal visibility. Thus the ground truth should be done at the same time, when the MSS sensor is collecting the data. Since the simultaneous observations are not always possible, it is preferable to estimate the optical thickness from the MSS data themselves.

7.6 General Models and Approximations

As we discussed in the previous section, two general models of atmospheric correction of terrestrial images are considered here. One applies to satellite data, i.e., the case where the remote sensor is located above the atmospheric

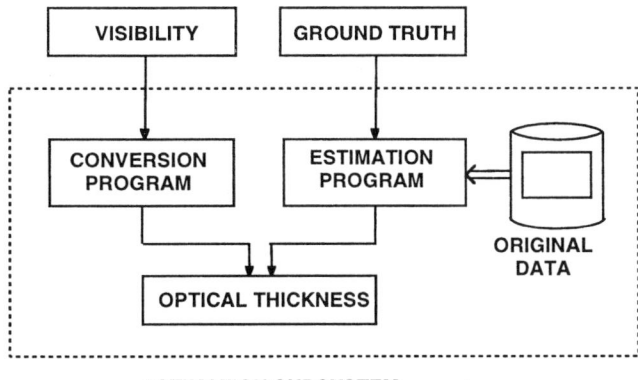

Figure 7.7. The block diagram of the estimation subsystem.

layer. The other is designed for airplane data, i.e., the case where the remote sensor is located in the atmosphere. These general models consider all multi-scattering and do not use the averaging method. Our model construction and approximation solution are based on the scattering theory, see Chapter 6, via the invariant imbedding concept see [10], [11] and [25].

7.6.1 General solution

The surface reflection at a point (x, y) on the earth is denoted by $K = K(x, y, \Omega, -\Omega')$. The matrix K is what we seek to estimate using remotely sensed data. The bidirectional reflection K represents the probability that a photon incident on the ground at point (x, y) in the direction $-\Omega'$ will reflect in the direction Ω within a solid angle, where $\pm\Omega = (\pm v, \phi)$ with v in the cosine of the polar angle measured from the normal to the top $0 \leq v \leq 1$ and ϕ the azimuthal angle $0 \leq \phi \leq 2\pi$. Above the ground is a layer of plane-parallel, inhomogeneous, anisotropic scattering atmosphere of optical thickness $h = w$. Associated with this atmosphere the reflections and transmissions are governed by the scattering matrix,

$$\mathfrak{S} = \mathfrak{S}(0, w) = \begin{pmatrix} t & \rho \\ r & \tau \end{pmatrix}. \tag{7.10}$$

Here $t = t(0, w; -\Omega, -\Omega')$ and $r = r(0, w; -\Omega, \Omega')$ are upward transmission and reflection operators for a layer extending from optical thickness $h = 0$ to $h = w$ and $\tau = \tau(0, w; \Omega, \Omega')$ and $\rho = \rho(0, w; -\Omega, \Omega')$ are downward transmission and reflection operators. The matrix $\mathfrak{S}$ is independent of (x, y), because our physical model for the atmosphere layer is plane-parallel. The domains and ranges of such operators are intensities of radiation and

$$\mathfrak{S}(0,w) \begin{pmatrix} I(w;x,y;-\Omega) \\ I(0;x,y;\Omega) \end{pmatrix} = \begin{pmatrix} I(0;x,y;-\Omega) \\ I(w;x,y;\Omega) \end{pmatrix}. \tag{7.11}$$

Sometimes for convenience $I(z;x,y;\pm\Omega)$ is represented by $I^\pm(z)$. $I^+(z)$, $0 \le z \le w$, denotes the downward intensity at optical thickness z above the point (x,y) on the ground in the direction $+\Omega$, and $I^-(z)$ signifies an upward intensity in the direction $-\Omega$.

The overall reflection at the surface of the atmosphere with a ground reflector as background is given by

$$P(K) = \rho + tK(E - rK)^{-1}\tau, \tag{7.12}$$

where E is the identity and the operator $(E - rK)^{-1}$ is assumed to exist and to be bounded. The reflected intensity at $h = w$ or the intensity received by the remote sensor is

$$I^-(w) = P(K)I^+(w); \tag{7.13}$$

see Figure 7.8. $I^+(w)$ is the intensity of radiation incident on the top of the atmospheric layer. If there were no atmospheric layer, then

$$I_0 = KI^+(w), \tag{7.14}$$

and $I^+(w) = I^+(0)$. Hence I_0 in equation (7.14) is the true ground radiation distribution while $I^-(w)$ in equation (7.13) is the blurred ground radiation distribution taken by a satellite remote sensor at or above the atmosphere. Their difference is due to the operators K and $P(K)$. The relation between them is expressed by equation (7.13). In equation (7.12), ρ is the reflection due to the atmospheric layer only while the term $(E - rK)^{-1}$ denotes the multiple scattering between the atmospheric layer and ground. To correct the atmospheric effect from the radiance data $I^-(w)$ in theory, it is only required to find the inverse of the operation P in equation (7.12). Since P maps K to $P(K)$, and P^{-1} recovers K from $P(K)$, i.e.,

$$KI^+(w) = P^{-1}[P(K)I^+(w)]. \tag{7.15}$$

Many experts in the field realize that, regardless of the model used, the essence of the problem is to solve the inverse problem. As it appears in equations (7.12) and (7.15), the inverse is not easy to find or to estimate. The first step is to find t, τ, ρ and r for a given optical thickness and a proper selection of the atmospheric model. To find such operators it is required to solve a set of nonlinear Riccati-type differential-integral equations, as we discussed in the previous chapters, and to use such results to find the inverse. It is again not easy because if one treats $(E - rK)^{-1}$ as a Neumann series, then the right-hand side of equation (7.12) involves an infinite number of terms of K. Even if one is successful in such an attempt, it is hardly a method which leads to a numerical solution. Furthermore, if one changes the atmospheric model, a new analysis must start from the beginning. Our method is presented in the next section with the intention of circumventing the above difficulties.

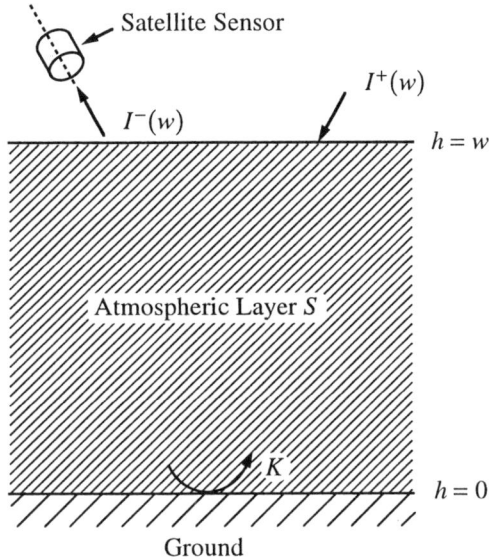

Figure 7.8. Cross sectional view of radiation received by a satellite sensor located above the atmospheric layer.

For the second mathematical model, the atmospheric layers are separated into two; one is above the airplane and the other is below, see Figure 7.9. These atmospheric layers are governed by scattering matrices $\mathfrak{S}_a = \mathfrak{S}(z, w)$ and $\mathfrak{S}_b = \mathfrak{S}(0, z)$. $\mathfrak{S}_a$ extends from optical thickness $h = z$ to $h = w$, while $\mathfrak{S}_b$ is from $h = 0$ to $h = z$. There are multiple scatterings in each layer, and interactions between $\mathfrak{S}_a$ and $\mathfrak{S}_b$, and the ground reflection K. It is necessary to analyze the so-called "internal radiation field" in a three-dimensional problem of radiative transfer with nonuniform background distribution. There is no reason to believe that an airplane will have a clearer image than that obtained by satellite, even if the airplane is much closer to the earth.

The scattering matrix $\mathfrak{S}_a = \mathfrak{S}(z, w)$ relates intensities at optical thickness $h = z$ and $h = w$ in the following manner:

$$\mathfrak{S}_a \left(\begin{array}{c} I^+(w) \\ I^-(z) \end{array} \right) = \left(\begin{array}{c} I^+(z) \\ I^-(w) \end{array} \right) \quad \text{with} \quad \mathfrak{S}_a = \left(\begin{array}{cc} t_1 & \rho_1 \\ r_1 & \tau_1 \end{array} \right) \tag{7.16}$$

and

$$I^-(w) = r_1 I^+(w) + \tau_1 I^-(z), \tag{7.17}$$

where $I^-(z)$ is the intensity measured on data taken by an airplane. To analyze the internal radiation problem, we consider a physical model, see Figure 7.10, which consists of only $\mathfrak{S}_b$ and ground reflection K. Then, as we discussed in the previous case, the overall reflection at the top of $\mathfrak{S}_b$ is

$$K_1 = \rho_2 + t_2 K (E - r_2 K)^{-1} \tau_2 = P_2(K), \quad \text{with} \quad \mathfrak{S}_b = \left(\begin{array}{cc} t_2 & \rho_2 \\ r_2 & \tau_2 \end{array} \right). \tag{7.18}$$

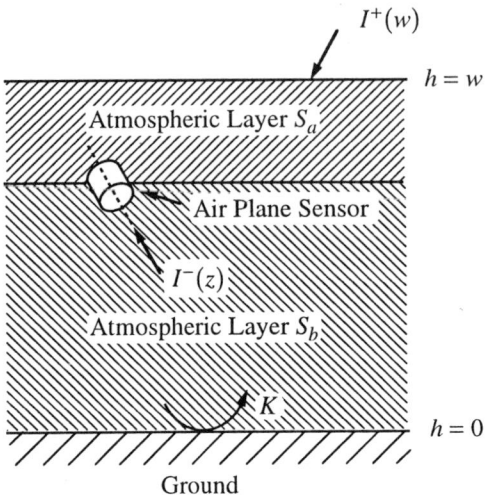

Figure 7.9. Cross section view of radiation received by an airplane sensor in the atmospheric layer.

From the reflection point of view, the physical model in Figure 7.9 can be replaced by that in Figure 7.11. That is, the atmospheric layer governed by $\mathfrak{S}_a$ is bounded below by a reflector K_1. The validity of this statement follows from invariant imbedding. Thus,

$$I^-(z) = K_1 I^+(z). \tag{7.19}$$

By equations (7.19) and (7.18),

$$I^-(z) = Q I^+(w) \tag{7.20}$$

where

$$Q = K_1 (E - r_1 K_1)^{-1} \tau_1. \tag{7.21}$$

For the airplane data correction problem, the true ground radiation distribution is given by I_0, as in equation (7.14). However, in order to obtain K, there are two inverse operations that must be solved; first one obtains K_1 from Q by equation (7.21) and then K from K_1 by equation (7.18). Both operations are equally as difficult as the inverse that occurs in equation (7.12), in theory. But $\mathfrak{S} = \mathfrak{S}(0, w)$ is more difficult to compute or to estimate than either $\mathfrak{S}_a$ or $\mathfrak{S}_b$, because of their differences in optical thickness. The problem involving multiple scattering in general is more complicated to handle with large values of optical thickness. In particular, the model is nonhomogeneous and anisotropic.

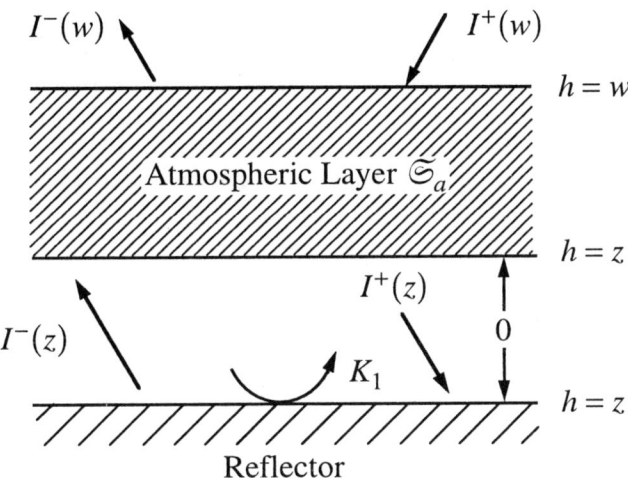

Figure 7.10. Cross section view of multi-scattering due to the atmospheric layer above the airplane.

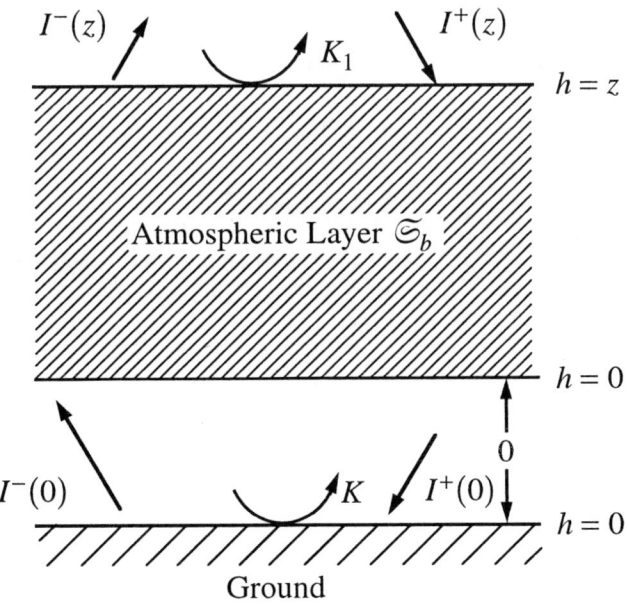

Figure 7.11. Cross section view of multi-scattering due to the atmospheric layer under the airplane.

7.6.2 Approximation methods

Our methods of solving the above inverse problems and finding approximate solutions suitable for a large computer are presented in this section. The reflection operators K, K_1, Q are imbedded in a larger system. We use the $*$-product of Chapter 6 and its linearization and the fact that for $\mathfrak{S}$ with very small optical thickness it can be approximated by its generating coefficients.

It is observed that, under the $*$-product,

$$\begin{pmatrix} -- & K \\ -- & -- \end{pmatrix} * \mathfrak{S}(0, w) = \begin{pmatrix} -- & P(K) \\ -- & -- \end{pmatrix}, \tag{7.22}$$

where $--$ denotes entries that are irrelevant, and $P(K)$ is as in equation (7.12). Let us partition $(0, w) = (0, w_1, w_2, \ldots, w_n = w)$ and set $\mathfrak{S}(w_j, w_{j+1}) = \mathfrak{S}_j$, $j = 1, 2, \ldots, p, \ldots n$ and $w_p = z$, then

$$\mathfrak{S}(0, w) = \mathfrak{S}_1 * \mathfrak{S}_2 \ldots * \mathfrak{S}_n; \tag{7.23}$$

since the $*$-product is difficult to compute, we shall use Mason's exchange rule or Redheffer's hat operation, see [25], if t is nonsingular,

$$\hat{\mathfrak{S}} : \mathfrak{S} \longrightarrow \hat{\mathfrak{S}} \quad \text{or} \quad \hat{\mathfrak{S}} = \begin{pmatrix} t^{-1} & -t^{-1}\rho \\ rt^{-1} & -rt^{-1}\rho + t \end{pmatrix} \tag{7.24}$$

then it is well known that

$$\hat{\mathfrak{S}}(0, w) = \hat{\mathfrak{S}}_1 \cdot \hat{\mathfrak{S}}_2 \cdots \hat{\mathfrak{S}}_n, \tag{7.25}$$

where $\cdot$ denotes the usual matrix multiplication. For n large, i.e., $w_{j+1} - w_j$ small, there is for each $\hat{\mathfrak{S}}_j$ an inverse. Therefore,

$$\hat{\mathfrak{S}}(0, w)^{-1} = \hat{\mathfrak{S}}_n^{-1} \cdot \hat{\mathfrak{S}}_{n-1}^{-1} \ldots \hat{\mathfrak{S}}^{-1}. \tag{7.26}$$

By the fact that the hat operation is nilpotent and it commutes with the inverse operation, we can find $\mathfrak{S}^{-1}(0; w)$ by one final hat operation:

$$\begin{pmatrix} -- & K \\ -- & -- \end{pmatrix} = (\hat{\mathfrak{S}}_n^{-1} \cdot \hat{\mathfrak{S}}_{n-1}^{-1} \ldots \hat{\mathfrak{S}}_1^{-1})\hat{} \cdot \begin{pmatrix} -- & P(K) \\ -- & -- \end{pmatrix}, \tag{7.27}$$

where $\hat{\mathfrak{S}}_j^{-1}$, $j = 1, 2, \ldots, n$, can be estimated by its coefficient, see below.

Since equation (7.18) is identical to equation (7.12) by replacing $S(0, w)$ by $\mathfrak{S}_b$ and K by K_1, we merely repeat the analysis; the result is

$$\begin{pmatrix} -- & K_1 \\ -- & 0 \end{pmatrix} = (\hat{\mathfrak{S}}_p^{-1} \cdot \hat{\mathfrak{S}}_{p-1}^{-1} \ldots \hat{\mathfrak{S}}_1^{-1})\hat{} \cdot \begin{pmatrix} -- & P_2(K) \\ -- & -- \end{pmatrix}. \tag{7.28}$$

To solve equation (7.20), we observe that

$$\begin{pmatrix} -- & K_1 \\ -- & -- \end{pmatrix} * \begin{pmatrix} E & 0 \\ r_1 & \tau_1 \end{pmatrix} = \begin{pmatrix} -- & Q \\ -- & -- \end{pmatrix}. \tag{7.29}$$

After computing the inverse under the $*$-product of the matrix

$$\begin{pmatrix} E & 0 \\ r_1 & \tau_1 \end{pmatrix}, \quad \text{and using} \quad \tilde{\mathfrak{S}} = \begin{pmatrix} t - \rho\tau^{-1}r & \rho\tau^{-1} \\ \tau^{-1} & -\tau^{-1}r \end{pmatrix}, \quad (7.30)$$

one obtains the matrix equation for the ground reflection matrix operator, K,

$$\begin{pmatrix} -- & K \\ -- & -- \end{pmatrix} = \begin{pmatrix} -- & Q \\ -- & -- \end{pmatrix} * \left[\begin{pmatrix} 0 & 0 \\ 0 & E \end{pmatrix} \cdot \tilde{\mathfrak{S}}_a \cdot \begin{pmatrix} 0 & E \\ E & 0 \end{pmatrix} \right.$$
$$\left. + \begin{pmatrix} E & 0 \\ 0 & 0 \end{pmatrix} \right]$$
$$(7.31)$$

where

$$\tilde{\mathfrak{S}}_a = \tilde{\mathfrak{S}}_{p+1} \cdot \tilde{\mathfrak{S}}_{p+2} \ldots \tilde{\mathfrak{S}}_n, \quad (7.32)$$

with $\tilde{\mathfrak{S}}_j = \tilde{\mathfrak{S}}(w_j, w_{j+1})$ and w_j the partition points used in equation (7.23). Again $\tilde{\mathfrak{S}}_a$ is obtained under the usual matrix multiplication, not under the more difficult $*$-product. Hence, K is solved if each $\tilde{\mathfrak{S}}_j$ in equation (7.32) can be computed.

The remaining section is devoted to estimating $\hat{\mathfrak{S}}_i$ and $\tilde{\mathfrak{S}}_j$, used in equations (7.28) and (7.32) for the radiative problem under consideration with arbitrary phase functions $p = p(w; \pm\Omega, \pm\Omega')$. These two parameters usually determine the physical characteristics of a nonhomogeneous atmospheric layer. Therefore, our method is rather general.

For $w_{j+1} - w_j = \Delta_j$ small, then

$$\mathfrak{S}_i = \mathfrak{S}(w_j, w_{j+1}) \doteq E + M(w_j)\Delta_j + 0(\Delta_j), \quad (7.33)$$

where $\doteq$ denotes estimation, $\|0(\Delta_j)\| \to 0$ as $\Delta_j \to 0$ and the coefficient

$$M(w_j) = \begin{pmatrix} a_j & b_j \\ c_j & d_j \end{pmatrix}. \quad (7.34)$$

For radiative transfer, M is decomposed into, see [22],

$$\begin{pmatrix} a_i & b_j \\ c_j & d_j \end{pmatrix} = \underset{\substack{\text{specular}\\\text{part}}}{\begin{pmatrix} p_j & 0 \\ 0 & q_j \end{pmatrix}} + \frac{1}{2\pi} \underset{\substack{\text{diffuse}\\\text{part}}}{\begin{pmatrix} P_j^{--} & P_j^{+-} \\ P_j^{-+} & P_j^{++} \end{pmatrix}}, \quad (7.35)$$

where the specular operators are given by

$$p_j I^-(w_j) = \int \exp(\Delta_j\mu_j)\delta(-\Omega, -\Omega')I(w_j; x, y; -\Omega')d\Omega'$$

and

$$(7.36a)$$

$$q_j I^+(w_j) = \int \exp(\Delta_j v_j)\delta(\Omega, \Omega')I(w_{j+1}; x, y; \Omega')d\Omega'$$

and the diffuse operators by

$$P_j^{\pm} I^-(w_j) = \int p(w_j, \pm\Omega, -\Omega')I(w_j; x, y; -\Omega')d\Omega'$$

and   (7.36b)

$$P_j^{\pm} I^+(w_j) = \int p(w_j; \pm\Omega, \Omega')I(w_{j+1}; x, y; \Omega')d\Omega'.$$

The integrations are all over the region $0 \leq \mu \leq 1$ and $0 \leq 1$ and $0 \leq \phi \leq 2\pi$. The above integral operators can be approximated by matrices A_i, B_i, etc., i.e.,

$$\begin{pmatrix} a_j & b_j \\ c_j & d_j \end{pmatrix} \doteq \begin{pmatrix} A_j & B_j \\ C_j & D_j \end{pmatrix}.$$ (7.37)

Then, under the hat operation, we obtain

$$\hat{S}_j = E + \hat{N}_j \Delta_j + \hat{M}_j \Delta_j^2, \qquad j = 1, 2, \dots, p, \dots, n,$$ (7.38)

where

$$\hat{N}_j = \begin{pmatrix} -A_j & -B_j \\ C_j & D_j \end{pmatrix} \quad \text{and} \quad \hat{M}_j = \begin{pmatrix} A_j^2 & A_j B_j \\ -C_j A_j & -C_j D_j \end{pmatrix}.$$

Similarly, we also obtain

$$\tilde{S}_j = E + \tilde{N}_j \Delta_j + \tilde{M}_j \Delta_j^2,$$ (7.39)

where

$$\tilde{N}_j = \begin{pmatrix} A_j & B_j \\ -C_j & -D_j \end{pmatrix} \quad \text{and} \quad \tilde{M}_j = \begin{pmatrix} -B_j C_j & -B_j D_j \\ C_j D_j & -D_j^2 \end{pmatrix}.$$

Finally, by substituting equations (7.38) and (7.39) into equations (7.25), (7.26), (7.31) and (7.32), we obtain the ground reflection matrix operator K for the satellite and for the airplane correction problems, respectively.

We have developed general models and solved the atmospheric correction problems. Our results are suitable for a large modern computer; they involve mostly matrix multiplications, and it is not necessary to determine transmission and reflection functions in advance in order to solve the inverse problems.

7.7 Results and Discussion

In this chapter we developed two correction methods and a general solution for the removal of atmospheric effects from remote sensing data. The mathematical models are based on the three-dimensional radiative transfer of a one-dimensional free atmosphere bounded by a horizontally non-uniform reflector. Two separate models are considered. One is the remote sensor located above the atmosphere, such as data taken by satellite. In the other the sensor is mounted on an airplane flying in the atmosphere. Atmospheric correction procedure for aircraft data is more complicated in comparison with that for atmospheric data, since the platform is located within the scattering medium.

Two correction methods are convolution and averaging. The convolution method is a single-reflection approximation. When it is applied to remote sensing in space, we are able to improve images of coastline and river. The averaging method is used for aircraft sensors in the atmosphere. We observe some good results in recognizing road and coastlines.

The general solution using the scattering matrix considers all multiple-scattering events. For satellite data correction, we first consider one-dimensional transmission and reflection operators in free space. Then we solve the inverse problem with an unknown uniform reflector. For aircraft data correction, the atmosphere is separated into two parts: one is above the airplane and the other is below. In this case we have to solve two inverse solutions. Approximation solutions suitable for implementation on digital computers are presented.

References

1.   G. W. Kattawar and G. N. Plass, "Radiance and Polarization of Multiple Scattering Light from Haze and Clouds," *Appl. Optics*, Vol. 7, 1968, pp. 1519–1527.

2.   R. E. Turner and M. N. Spencer, "Atmospheric Model for Correction of Spacecraft Data," in *Proceedings of Eighth International Symposium of Remote Sensing of Environment*, University of Michigan, Ann Arbor, Michigan, 1972, pp. 895–934.

3.   J. F. Potter and M. Shelton, "Effect of Atmospheric Classification of ERTS 1," in *Proceedings of Ninth International Symposium on Remote Sensing of Environment*, University of Michigan, Ann Arbor, Michigan, 1974, pp. 865–874.

4.   M. Griggs, "Determination of Aerosol Content in Atmospheres from ERTS 1 Data," in *Proceedings of Ninth International Symposium*, Ann Arbor, Michigan, 1974, pp. 471–481.

5.   J. F. Potter and M. A. Mendlowitz, "On the Determination of Haze Levels from LANDSAT Data," in *Proceedings of the Tenth International Symposium on Remote Sensing of Environment*, University of Michigan, Ann Arbor, Michigan, 1975, pp. 695–703.

6.   T. Takashima, "A New Approach of the Adding Method for the Computation of Emergent Radiation in an Inhomogeneous, Plane-parallel Planetary Atmosphere," *Astrophys. Space Science*, Vol. 36, 1975, pp. 319–328.

7.   R. S. Fraser, "Interaction Mechanisms within the Atmosphere," in *Manual of Remote Sensing*, R. G. Reaves, editor, American Association of Photogrammetry, 1975, pp. 181–233.

8.   S. Ueno and S. Mukai, "Atmospheric Effects on Remotely Sensed Data from Space," in *Proceedings of IFAC Symposium on Environmental Systems Planning, Design and Control*, H. Akashi, editor, Pergamon Press, Oxford, 1977, pp. 423–428.

9.   S. Ueno, "Invariant Imbedding and Diffuse Reflection of Radiation by Multidimensional Slabs," *Memoires of Kanazawa Institute of Technology*, Vol. 6, 1976, pp. 1–41.

10.   A. P. Wang, "Correction of Atmospheric Effects on Remote Sensing," *Math. Modeling*, Vol. 9, No. 2, 1987, pp. 117–124.

11.   A. P. Wang, "An Inverse Problem of Reflection," *Comp. Math. Appl.*, Vol. 25, No. 9, 1993, pp. 59–65.

12.   A. P. Odell and J. A. Weinman, "The Effect of Atmospheric Haze on Images of the Earth's Surface," *J. Geophys. Research*, Vol. 80, 1975, pp. 5035–5040.

13.   R. E. Turner, "Signature Variations Due to Atmospheric Effects," in *Proceedings of the Tenth International Symposium on Remote Sensing of Environment*, University of Michigan, Ann Arbor, Michigan, 1975, pp. 671–682.

14.   Y. Haba, Y. Kawata, and Y. Terashita, "Atmospheric Effects on the Classification of Remotely Sensed Earth's Imageries," in *Proceedings of IFAC Symposium on Environmental Systems Planning, Design and Control*, H. Akashi, editor, Pergamon Press, Oxford, 1977, pp. 409–414.

15.   S. Ueno, "Contrast Transmittance at the Top of an Atmosphere Bounded by a Horizontally Non-uniform Diffuse Reflector," in *Proceedings of the Tenth Lunar and Planetary Symposium*, Institute of Space and Aeronautical Science, University of Tokyo, 1977, pp. 166–170.

16.   S. Ueno, Y. Haba, Y. Kawata, T. Kusaka and Y. Terashita, "The Atmospheric Blurring Effect, and Their Correction in Airborne Sensor and LAND-SAT MSS Data," in *Proceedings of the Twelfth International Symposium on Remote Sensing of Environment*, April 20–26, Manila, Philippines, 1978, pp. 1241–1257.

17.   Y. Kawata, Y. Haba, T. Kusaka, Y. Terashita, and S. Ueno, "Atmospheric Effects and Their Correction in Air-borne Sensor and LANDSAT MSS Data," in *Proceedings of the Twelfth International Symposium on Remote Sensing of Environment*, April 20–26, Manila, Philippines, 1978, pp. 1241–1257, ERIM, Ann Arbor.

18.   T. Kusaka, Y. Haba, Y. Kawata, Y. Terashita and S. Ueno, "Removal of the Atmospheric Blurring from Remotely Sensed Earth Imagery," in *Proceedings of the 4th International Joint Conference on Pattern Recognition*, 1978, Kyoto, Japan, pp. 931–935.

19.   Y. Kawata, Y. Haba, T. Kusaka and S. Ueno, "Atmospheric Correction for the Earth Imagery by Air-Borne Sensor, in 1978," *Proceedings of the International Conference on Cybernetics and Society*, Tokyo-Kyoto, Japan, Nov. 3–7, 28CH 1306-0 SMC, pp. 667–671.

20.   S. Ueno, "Invariant Imbedding in a Two-dimensional Slab Illuminated by Parallel Rays," in *Proc. Intl. Association of Meteorology and Atmospheric Physics (IUGG)*, 1979, pp. 10–15.

21.   Y. Kawata, Y. Haba and S. Ueno, "The System of Correction g Remote Sensed Earth's Image for Atmospheric Effects" *Proc. of 13th Int. Symp. on Remote Sensing of Environ.*, ERIM, Ann Arbor, Michigan, April 23–27, 1979, pp. 1883–1894.

22.   L. Elterman, "UV, Visible and IR Attenuation for Altitude to 50 km," *Rept. AFGRL-68-0153*, AFCRL, Bedford, Mass., 1968.

23.   D. Deirmendjan, *Electromagnetic Scattering on Spherical Polydispersions*, American Elsevier, New York, 1969.

24.   J. E. Hansen and L. D. Travis, "Light Scattering in Planetary Atmospheres," *Space Science Review*, Vol. 16, 1974, pp. 527–610.

25.   R. Redheffer, "On the Relations of Transmission Line Theory to Scattering and Transfer," *J. of Math. Phys.*, Vol. 41, No. 1, 1962, pp. 1–41.

8. Topographic Effects in Terrestrial Remote Sensing

The surface radiance distribution measured in remote sensing from space differs from its value at the ground because of the presence of the atmosphere [1]–[9]. The removal of atmospheric effects from space-based images of the earth improves the accuracy of the classification of the ground objects. A further correction is needed for mountainous terrain. The topographic effect refers to the obscuration of terrestrial information due to the effect of terrain on the reflectance at the surface. The radiative transfer problem is very difficult to solve precisely.

In this chapter it is shown how to effectively determine and remove atmospheric and topographic effects from the pixels of space-based images. Practical approximations to the solution of the radiative transfer problem are suggested. The bidirectional reflectance law can be treated by "model rendering," similar to that used in computer graphics. Removal of the topographic and atmospheric effects from satellite data enables an improved retrieval of terrestrial information, such as "hidden" lakes.

8.1 Introduction

In previous chapters we have developed increasingly realistic models of terrestrial radiative transfer. We have shown how to remove the effects of atmospheric scattering from the data obtained from satellites and aircraft for the determination of ground characteristics. However, the topographic effect, due to the reflection of atmospheric radiation by the terrain on the spectral response from space, is a complex and difficult problem. Here, we provide an approach based on approximating the three-dimensional problem by a one-dimensional one and the use of model rendering.

8.2 Flat Terrain

8.2.1 Three-dimensional model

Consider a three-dimensional model consisting of a free atmosphere bounded by a flat reflecting surface with nonuniform albedo distribution. Let us measure distances in optical units. The top z of a plane-parallel, vertically inhomogeneous, anisotropically scattering atmosphere is uniformly and monodirectionally illuminated by parallel rays of constant net flux πF per unit area normal to the direction of propagation, Ω_0. The variable attenuation and scattering coefficients are given by α and σ, respectively. It is customary to set the albedo for single scattering equal to the ratio of the scattering and attenuation coefficients,

$$\lambda = \sigma/\alpha. \tag{8.1}$$

The intensity of radiation going in the direction Ω from the level t ($0 \leq t \leq z$) at horizontal rectangular coordinates (x, y) is denoted by $I(t, x, y; \Omega)$. In the above Ω stands for $(\theta = \cos^{-1} v, \phi)$, where θ is a polar angle measured from the upward normal and ϕ is an azimuthal angle with respect to an arbitrary horizontal axis. The level-dependent phase function is expressed by $p(z; \Omega, \Omega')$.

8.2.2 Bidirectional reflectance

The intensity of the upwelling radiation at the bottom in the direction Ω can be related to the downwelling radiation as follows,

$$I(0, x, y, \Omega) = v^{-1} \int_{2\pi} k(x, y, \Omega, \Omega') I(0, x, y, -\Omega') v' d\Omega', \tag{8.2}$$

where $k(x, y, \Omega, \Omega')$ is the probability that a photon incident on the bottom $(0, x, y)$ in the direction $-\Omega'$ will be reflected from it in the upward direction Ω within an elementary solid angle. The total intensity of downwelling radiation is the sum of the directly and diffusely transmitted intensities,

$$I(0, x, y, -\Omega') = \pi F \delta(\Omega' - \Omega) \exp(-z/v') + I^*(0, x, y, -\Omega'), \tag{8.3}$$

where δ is the Dirac delta function and $I^*(0, x, y, -\Omega')$ is the diffusely transmitted intensity.

The bidirectional reflection law for reflection at the bottom surface is expressed through the function $k(x, y, \Omega, \Omega')$. For example, in the case of isotropic reflection,

$$k(x, y, \Omega, \Omega') = v A(x, y)/\pi, \tag{8.4}$$

where $A(x, y)$ is the horizontally inhomogeneous albedo of the Lambert reflecting surface and represents the ratio of the total energy reflected by the surface to the incident energy.

8.2.3 Three-dimensional scattering function

Equation (8.3) is a boundary condition that must be satisfied when the transport equation is solved for internal intensity. We shall avoid this unstable problem by addressing the three-dimensional scattering function S which yields the emergent diffuse intensity at the top, where Ω here stands for upwelling directions only,

$$I^*(z, x, y, \Omega) = FS(z, x, y, \Omega, \Omega_0)/4v. \tag{8.5}$$

This S function satisfies the integro-differential Eq. (4.48) derived in Section 6.6. When the bottom surface reflects isotropically according to Lambert's law,

$$I^*(0, x, y, \Omega) = A(x, y)\pi^{-1} \int_{2\pi} I(0, x, y, -\Omega')v'd\Omega', \tag{8.6}$$

then the initial condition on S is

$$S(0, x, y, \Omega, \Omega_0) = 4A(x, y)uv. \tag{8.7}$$

8.3 Rugged Terrain

8.3.1 Model of rugged terrain

In general, the effect of rugged terrain on atmospheric radiation fields should be determined in accordance with the bidirectional reflectance law at each target pixel of the image. The bidirectional spectral reflectance law depends on the terrain elevation and position, polar and azimuthal angles of the normal to the surface of the terrain in a given pixel, and the angles of incidence and reflection in the pixel.

To evaluate this effect exactly is difficult, so we make an approximation in which we ignore the variations of ground reflection due to the inclination of the target, and we approximate the scattering and transmission functions by the respective "equivalent" functions which only depend on vertical inhomogeneities, and not on x and y variations. Such scattering and transmission functions have been evaluated based on Elterman's model, see Chapter 7.

8.3.2 Target ground albedo

The target ground albedo is approximated [1],

$$A_t = (I_{\text{obs}} - FS(z, \Omega, \Omega_0)/4v)/X, \tag{8.8}$$

where

$$X = \cos\Theta F\{\exp[-z/v - z/u] + (4\pi)^{-1}\exp(-z/v)\int_{2\pi} T(Z,\Omega',\Omega_0)d\Omega'\}.$$
(8.9)

In these equations, I_{obs} is the observed radiance at the top and Θ is the angle between the solar incident direction and the local surface normal. Instead of using the three-dimensional S and T functions, we approximate them by the one-dimensional (dependence on only the vertical coordinate) scattering and transmission functions known through solution of the invariant imbedding equations of Section 6.1. The expression for Θ is given by

$$\cos\Theta = \cos\theta_0\cos\theta_n + \sin\theta_0\sin\theta_n\cos(\phi_n - \phi_0),$$
(8.10)

where θ_0 is the solar zenith angle, θ_n is the zenith angle of the normal to the target surface, ϕ_0 is the solar azimuthal angle, and ϕ_n is the direction of the maximum slope at the target. Equation (8.8) removes the atmospheric and topographic effects from Landsat data over rugged terrain. The reflectance of backgrounds surrounding the target, the "adjacent effect" of the background albedos, has not been included.

If we denote the tilt angle of the target surface as a, then we have $\phi_n = a$. Since the values of a and ϕ_n can be computed from the elevation data for a given terrain, we can obtain the values of Θ at each target location. Note that in Eq. (8.10) the explicit dependence on the height of the target does not appear. In a flat terrain, we have $\cos\Theta = u$ because $a = 0$. The negative value of $\cos\Theta$ corresponds to the case where no direct solar illumination is available at the target point.

8.3.3 Computational results

Equations (8.8) and (8.9) have produced improved shaded relief images in the classification ground albedo map from the Landsat computer compatible tape for the Kanazawa area of Japan [1], [2]. This technique provides a first, but practical, approximation for qualitative analysis of image data. See also Section 8.5.

8.4 Model Rendering

8.4.1 Introduction

The correction of terrestrial images obtained from satellites depends on the techniques employed to process the information. For each target pixel, the atmospheric effect should be estimated taking into account the bidirectional reflectance of the terrain. The generation of images through application of physical principles of radiative transfer to models of objects and media is called "model rendering" [2], [10]. The aim is to compute the intensity of the radiation coming from various directions and registering in the pixels of an image plane.

8.4.2 Model rendering integral equation

We consider the energy in a radiation field due to emission of light, transmission through a nonhomogeneous nonscattering medium, and multiple reflections at various surfaces in the medium. Note that the focus is not on multiple scattering of photons in the medium, but instead on multiple reflections at surfaces. It is assumed that phase, frequency and other temporal aspects can be neglected.

We introduce the functions

$i(\theta, \phi)$ = intensity of radiation at a point x travelling in a direction whose polar and azimuthal angles are θ and ϕ,

γ = distance between points x' and x,

$\rho(x, x', x'')$ = fraction of light travelling from a surface element at x'' and arriving at a surface element at x that is scattered by a surface element at x',

$e(x, x')$ = energy per unit time per unit area at x due to energy emitted at x',

$g(x, x')$ = geometrical factor that takes into account intervisibility between points x and x' as well as the spreading of energy from a point source:

         = 0, if there is no clear line of sight between x and x',

         = $1/r^2$, otherwise, where r is the distance between x and x'.

The function $I = I(x, x')$ is introduced as a measure of the radiation at x due to radiation coming from x',

$$I(x, x') = i(\theta', \phi') \cos \theta \cos \theta' / r^2, \qquad (8.11)$$

and it satisfies the integral equation,

$$I(x, x') = g(x, x') \left\{ e(x, x') + \int_S \rho(x, x', x'') I(x', x'') dx'' \right\}. \qquad (8.12)$$

The integral is over S, all the surfaces of the objects in the problem. Kajiya calls I the transport intensity. The quantity I has units of energy per unit time per unit area of source per unit area of the scatterer. If the medium between surfaces participates in the scattering, then the intensity of radiation satisfies an integro-differential equation, as we have seen before.

8.4.3 Approximate solution

To obtain the exact computational solution of this integral equation is difficult. Methods have been developed for approximate solutions. These include approximating the integral by a sum and solving a large system of linear algebraic equations; the difficulty is that the equations may be ill-conditioned and small changes in coefficients can produce large changes in solutions. Alternatively, one could approximate the solution by infinite series or by orders of scattering. There is also the possibility of simulation of ray tracing, by one reflection, two, three, etc.

In the case of a diffuse reflector, the surfaces have no angular dependence on the bidirectional reflectance function, and the following radiosity approximation is useful. The radiosity $B(x')$ is defined to be the integral over the hemisphere

$$dB(x') = dx' \int I(\theta', \phi) \cos\theta' d\Omega \qquad (8.13)$$

and the radiosity approximation is the integral over the surface

$$dB(x') = dx' \int I(x, x')dx. \qquad (8.14)$$

The radiosity is determined by computing the total integrated intensity. This is an intensive computation which involves numerous intervisibility calculations. The integral equation was solved by a Monte Carlo method with multidimensional sequential sampling by Kajiya [10] whose results were compared with a ray tracing approximation.

8.5 Topographic and atmospheric correction of satellite data

8.5.1 Topographic correction

The shadows due to the topography of terrain causes a pronounced effect on Landsat near-infrared radiance data [1]. This is found in the estimation of ground albedo from Landsat image data in band 7 taken on 23 October 1979 of the wooded lake area of Kanazawa, Japan. The ground albedo cross-section along the 64th line is computed with Eq. (8.8) using smoothed terrain data. The smoothed terrain database is generated by first digitizing the geographic map of the area with sampling intervals of 50 m. The Landsat-measured radiances are integrated over a reflected area of 57 m by 57 m, the terrain data are interpolated into intermediate terrain data with grid intervals of 25 m, then to new data with 57 m intervals using the cubic convolution method and nine neighboring points. This smoothed terrain database is used to determine the value of γ, the cosine of the angle between in incident radiation and the local normal to the surface for each pixel.

8.5.2 Estimation of ground albedo

A histogram analysis of the frequency of estimated albedos on line 64 of the image shows that the mean albedo is 0.33. The dispersion of estimates has a

standard deviation of 0.087, lower than that of the original data. The shape
of the distribution is close to being normal, which seems to help validate this
approach.

The estimated ground albedo of about 0.3 is consistent with the category
of deciduous forest. The gray image determined from the ground albedo estimates reveals a lake which was not evident in the original image. Although
many shadow regions are eliminated, those of the deep valleys remain. While
this study assumed a Lambertian surface, other reflectance laws can and
should be considered to improve the retrieval of information from satellite
data. Please see Ref. [1] for image data and results.

8.6 Discussion

In this chapter on the effects of terrain on reflectance, we have studied the
bidirectional reflectance law, the three-dimensional scattering function, the
target ground albedo, and model rendering. Finally we have presented results
of topographic and atmospheric correction of Landsat image data.

References

1.   Y. Kawata, S. Ueno and T. Kusaka, "Radiometric Correction for Atmospheric and Topographic Effects on Landsat MSS Image," *Int. J. Remote Sensing*, Vol. 9, 1988, pp. 729–748.

2.   H. H. Kagiwada, S. Ueno and Y. Kawata, "Simulation of Atmospheric and Topographic Effects in Remote Sensing from Space," *Internat. Geoscience and Remote Sensing Symposium*, May, 1990, College Park, MD.

3.   H. H. Kagiwada, S. Ueno and Y. Kawata, "Approximation of Topographic Effect on Atmospheric Correction in Rugged Terrain," in *Proc. of IGARSS'91*, Helsinki, June 3-6, 1991, pp. 665–668.

4.   Y. Kawata, A. Hatakeyama, T. Kuska and S. Ueno, "The Evaluation of Various Radiance Components in Mountainous Terrain," in *Proc. of IGARSS'92*, Houston, Texas, USA, May 26-29, 1992, Vol. 2, 1228–1230.

5.   H. Kagiwada, S. Ueno and Y. Kawata, "Analytical Approximation of Atmospheric Correction in Rugged Terrain," in *Proc. of ISPRS'92*, Washington, USA, Aug. 2-14, 1992, pp. 340–345.

6.   S. Ueno, O.I. Smokty and Y. Kawata, "Deblurring of Satellite Images in Flat Terrain Via Analysis," in *Proc. of IGARSS'93*, Tokyo, Japan, Aug. 18-21, 1993, pp. 404–406.

7.   S. Ueno, Y. Kawata and T. Takashima, "Topographic Effect on Atmospheric Correction in Rugged Terrain," in *Current Problems in Atmospheric Radiation*, S. Keevallili and O. Karner, Eds., Deepak Press, pp. 432–435.

8.   A. P. Wang, "Correction of Atmospheric Effects in Remote Sensing," *Math. Modeling, An International Journal*, Vol. 133, No. 2, 1988, pp. 383–394.

9.   K. Ya. Kondratyev, *The Atmospheric Effect in the Investigation of Natural Resources Coming from the Cosmos*, (in Russian), Moscow, Machinostroenie, 1985.

10.   I. T. Kajiya, "The Rendering Equation," *Computer Graphics*, Vol. 20, SIGGRAPH'86, 1986, pp. 143–150.

9. Searchlight Problem

The searchlight problem is concerned with the target reflection from a point source at the top of the atmospheric layer. In this chapter we develop methods to recover the true target reflection from the observed data which consists of the background light and the multiple scattering in the atmospheric layer. After the construction of the mathematical model of initial value problems based on invariant imbedding, we also establish approximation solutions and simulation results.

9.1 Introduction

In this chapter we discuss the illumination at the top of an atmospheric layer by a narrow collimated beam of a given frequency such as a searchlight or a laser beam. The atmospheric layer is a plane-parallel atmospheric medium with anisotropic scattering and is bounded below by a reflective earth surface. The top of the atmospheric layer is also illuminated by a low intensity uniform light, called the *background light*. Therefore the reflected light on the earth's surface due to the background light has a relative low intensity.

Our main result is the recovery of the true target reflection from the observed data at the top of the atmospheric layer. The observed data consists of the reflections of the searchlight and the background light. We consider all multiple-scattering in the atmospheric layer and between the atmospheric layer and the earth's surface.

For a general review of the history of work done on searchlights, see references [1–6]. The work presented here is rather general; see [7–9]. The earth's surface is illuminated by the searchlight on a small area called the *target*. The remaining area has low reflected intensities assumed to be almost uniform.

With the method of invariant imbedding and the scattering theorem discussed in the previous chapters, we construct a mathematical model of a searchlight. The model is that of three-dimensional radiative transfer, i.e., the specific intensity depends on (x, y, z) in a Cartesian coordinate system. For convenience, the intensities are separated into specular and diffuse parts. The searchlight produces mostly the specular part and the background the diffuse part. We first solve the reflection due to background light in *free space*, a one-dimensional problem. The overall reflected light is three-dimensional. Therefore, our answer can be built on the well-known results of one-dimensional radiative transfer. Besides the exact solution in the form of a series, we discuss the convergence and give first and second order of approximations. Finally, a numerical simulation result is given via Monte Carlo method.

9.2 Basic Equations

In this section we treat the searchlight problem based on a three-dimensional radiative transfer theory at a given frequency. The plane-parallel medium with anisotropic scattering is bounded by a flat reflecting surface with inhomogeneous albedo. The intensity is denoted by $I = I(z, x, y; \Omega)$, where $0 \leq z \leq z_1$, the optical thickness, and $-\infty \leq x, y < \infty$. Also $\Omega = (\theta, \phi)$, where $\theta = \cos^{-1} v$ is a polar angle measured from the outward normal at the top of the atmospheric layer and ϕ is an azimuthal angle. The equation of transfer appropriate to the present case takes the form

$$\Omega \cdot \nabla I(z, x, y; \Omega) + \alpha(z)I = \frac{\sigma}{4\pi} \int_{4\pi} p(z; \Omega, \Omega')I(z, x, y; \Omega')d\Omega' \qquad (9.1)$$

where $\Omega \cdot \nabla I$ is the directional derivative of the intensity in Cartesian coordinates, α the attenuation coefficient, σ the scattering coefficient, and p the depth-dependent phase function. The first term on the left hand side of eq. (9.1) is the variation of intensities in the direction Ω. The second term is the attenuation of the intensity due to direct transmission, i.e., the specular part. The right hand side of eq. (9.1) is the sum of all intensities $I(z, x, y, \Omega')$ scattered in the direction Ω, i.e., the diffused part. Computing the directional derivative, the equation of transfer takes the form

$$\cos \theta \frac{\partial I}{\partial z} + \sin \theta \cos \phi \frac{\partial I}{\partial x} + \sin \theta \sin \phi \frac{\partial I}{\partial y} + \alpha(z)I(z, x, y; \Omega)$$
$$= \frac{\sigma(z)}{4\pi} \int_{4\pi} p(z; \Omega, \Omega')I(z, x, y; \Omega')d\Omega'. \qquad (9.2)$$

Eq. (9.2) should be solved subject to the boundary conditions

$$I(z_1, x, y; -\Omega) = \pi[F\delta(\Omega - \Omega_0)\delta(x)\delta(y)] + I_b(z_1, x, y; -\Omega), \qquad (9.3)$$

and

$$I(0, x, y; \Omega) = \frac{1}{v} \int_{2\pi} k(\Omega, \Omega')I(0, x, y; -\Omega')v'd\Omega', \qquad (9.4)$$

where $\delta(\Omega - \Omega_0) = \delta(\theta - \theta_0)\delta(\phi - \phi_0)$ is the Dirac delta function, Ω_0 is the direction of incident intensity, and the bidirectional reflection kernel $k(\Omega, \Omega')$ represents the probability that a photon incident on the bottom of the atmospheric layer $(0, x, y)$ in the direction Ω' will be reflected in the upward direction Ω within an elementary solid angle. I_b is the background light. In the case of *isotropic reflection* we put

$$k(x, y; \Omega, \Omega') = \frac{vA(x, y)}{\pi}, \qquad (9.5)$$

i.e., k is independent of Ω and Ω' where $A(x, y)$ is the horizontally inhomogeneous albedo of the reflecting surface, e.g., in accordance with Lambert's law. In this case Eq. (9.4) becomes

$$I(0, x, y; \Omega) = \frac{A(x, y)}{\pi} \int_{2\pi} I(0, x, y; -\Omega') v' d\Omega'. \qquad (9.6)$$

In Eq. (9.6), $I(0, x, y; -\Omega')$ consists of the directly and diffused downwards transmitted intensities as below

$$I(0, x, y; -\Omega') = \pi \delta(\Omega - \Omega')[F\delta(x)\delta(y)]e^{-z_1/v'} + I^*(0, x, y; -\Omega'). \qquad (9.7)$$

In Eq. (9.7), z_1 denotes the total optical thickness and I^* represents the diffuse radiation field under consideration.

9.3 Equation of Transfer

Equation (9.2) is a local differential equation of intensity $I(z, x, y; \Omega)$ subject to two boundaries (9.3) and (9.4). Equation (9.3) denotes the initial incident intensity at $z = z_1$. To construct an initial value problem, we repeat the invariant imbedding procedures used in the previous chapters. A thin layer Δz is added to the upper boundary and we take the limit of the reflection operator $S(z, x, y; \Omega, \Omega_0)$ as $\Delta z \to 0$. Recall that

$$S(z, x, y; \Omega, \Omega_0) = 2vR(z, x, y; \Omega, \Omega_0) \qquad (9.8)$$

and

$$
\begin{aligned}
I(z, x, y, +v, \phi) &= \frac{1}{2\pi} \int_{-\infty}^{\infty} \int_{-\infty}^{\infty} \int_0^1 \int_0^{2\pi} R(z, x - x', y - y'; v, \phi; v', \phi) \\
&\quad \cdot I(z, x, y, -v', \phi') dx' dy' dv' d\phi',
\end{aligned} \qquad (9.9)
$$

for a slab of optical thickness z in a free space. The result is

$$
\begin{aligned}
&\frac{\partial S}{\partial z}(z, x, y; v, \phi; v_0, \phi_0) + (\tan\theta \cos\phi + \tan\theta_0 \cos\phi_0)\frac{\partial S}{\partial x} \\
&+ (\tan\theta \sin\phi + \tan\theta_0 \sin\phi_0)\frac{\partial S}{\partial y} + \left(\frac{1}{v} + \frac{1}{v_0}\right)\alpha(z)S \\
&= \sigma(z)\left[p(z_1; v, \phi; -v_0, \phi_0)\delta(x)\delta(y)\right. \\
&+ \frac{1}{4\pi} \int_0^1 \int_0^{2\pi} S(z, x, y; v, \phi; v', \phi')p(z; -v', \phi'; -v_0, \phi_0)\frac{dv'}{v'}d\phi'
\end{aligned} \qquad (9.10)
$$

$$+ \frac{1}{4\pi} \int_0^1 \int_0^{2\pi} p(z; v, \phi; v'', \phi'') S(z, x, y; v'', \phi''; v_0, \phi_0) \frac{dv''}{v''} d\phi''$$

$$+ \frac{1}{16\pi^2} \int_{-\infty}^{\infty} \int_{-\infty}^{\infty} \int_0^1 \int_0^{2\pi} \int_0^1 \int_0^{2\pi} S(z, x - x', y - y'; v, \phi; v', \phi')$$

$$\cdot \quad p(z; -v', \phi'; v'', \phi'') S(z, x', y'; v'', \phi''; v_0, \phi_0) dx' dy' \frac{dv'}{v'} d\phi' \frac{dv''}{v''} d\phi'' \Bigg].$$

Equation (9.10) is the general three-dimensional equation of transfer. It was stated in Chapter 6, equation (4.48). For the searchlight under consideration, equation (9.10) is subject to the initial condition

$$S(0, x, y; \Omega, \Omega_0) = \overline{A}(x, y) v v_0 \delta(x) \delta(y) \tag{9.11}$$

where

$$\overline{A}(x, y) = \begin{cases} A(x, y) & \text{when } (x, y) \subset \text{small target area} \\ \text{constant} & \text{otherwise} \end{cases} \tag{9.12}$$

In case S is independent of y, then $\frac{\partial S}{\partial y} = 0$ and eq. (9.10) reduces to

$$\frac{\partial S}{\partial z}(z, x; \Omega, \Omega_0) + (\tan\theta\cos\phi + \tan\theta\cos\phi_0)\frac{\partial S}{\partial x} + \alpha(z)\left(\frac{1}{v} + \frac{1}{v_0}\right)S$$

$$= \sigma(z)[p(z; \Omega, -\Omega_0)\delta(x)$$

$$+ \frac{1}{4\pi} \int_0^{2\pi} S(z, x; \Omega, \Omega'') p(z; -\Omega'', -\Omega_0) \frac{\partial\Omega''}{v''} \tag{9.13}$$

$$+ \frac{1}{4\pi} \int_0^{2\pi} p(z; \Omega, \Omega') S(z, x, y; \Omega', \Omega_0) \frac{d\Omega'}{v'}$$

$$+ \frac{1}{16\pi^2} \int_{-\infty}^{\infty} \int_0^{2\pi} S(z, x - x'; \Omega, \Omega') p(z; -\Omega', \Omega'') S(z, x'; \Omega'', \Omega_0) dx'$$

$$\cdot \quad \frac{d\Omega'}{v'} d\frac{\Omega''}{v''}.$$

Equation (9.13) is the equation of transfer of a two-dimensional searchlight problem. This equation is subject to the initial condition

$$S(0, x; \Omega, \Omega_0) = 4\overline{A}(x) v v_0 \delta(x). \tag{9.14}$$

Equations (9.13) and (9.14) are a useful model for solving problems of the earth's surface consisting of land and ocean.

It should be mentioned that the solution of (9.13) with the initial condition (9.14) is invariant with respect to input at the top of the atmospheric layer. In this case, as stated in the introduction, the atmospheric layer is also illuminated by a low uniform background light, say $I_b = I_b(z_1)\delta(\Omega - \Omega_b)$; then the total incident intensity is

$$I_{\text{inc}} = I_S + I_b \tag{9.15}$$

$$I(z_1, x, y; \Omega) \;=\; FS(z_1, x, y; \Omega, \Omega_0) \tag{9.16}$$

$$+ \;\; \frac{1}{v} \int_{-\infty}^{\infty} \int_{\infty}^{\infty} S(z_1, x - x', y - y'; \Omega, \Omega_b) I_b(z, x', y') dx' dy'.$$

In general for the searchlight problem the incident intensities at the top of the atmospheric layer consist of radiation from the searchlight and from the background light. The background light radiation usually has a wide frequency spectrum while the searchlight has a narrow spectrum. The ideal frequency for a searchlight is concentrated at 2.1–2.3 or 3.7–4.0 μm where the transmittance of the atmosphere is at its maximum. Within this narrow band of frequency, the searchlight intensity is much stronger than that of the background. Equations (9.10) and (9.13) should be used for all parameters at these narrow bands.

The following sections are devoted to developing approximate solutions for the searchlight problem.

9.4 Asymptotic Solutions

We stated that the exact solution is very difficult to obtain. In this section, we discuss the asymptotic and approximate solutions. The asymptotic solution is based on the scattering theory discussed in Chapter 6, and it converges to the exact solution. The system is conservation and the ground reflection is dissipative. Knowing the convergent property [10], it is meaningful to compute the first n terms of solutions. Those n terms correspond to the exact nth order of multiplication. This is called nth order approximation.

In case the initial condition given by eq. (9.11) is replaced by $S(0, x, y; \Omega, \Omega_0) \equiv 0$, i.e., there is no boundary reflection at the bottom of the atmospheric layer. We denote the solution of (9.10) subjected to this new boundary as $\rho = \rho(z, x, y, \Omega)$. Under the assumption that the atmospheric layer is a plane-parallel medium which is independent of x and y, so is the initial condition. Hence, it suffices to consider $\partial \rho / \partial x = 0$ and $\partial \rho / \partial y = 0$ or ρ is a function of (z, Ω, Ω'). Eq. (9.10) reduces to a reflection equation of Chandrasekhar's type [2] and the solution ρ is well-known.

For convenience, we express equation (9.4) as

$$I(0, x, y; \Omega) = KI(0, x, y; -\Omega') \tag{9.17}$$

where the integral operator $K = K(\Omega, \Omega')$, associated with kernel $k(\Omega, \Omega')$. The relation between S and ρ is given by

$$S = S(K) = \rho + tK[E - rK]^{-1}\tau, \tag{9.18}$$

$S(K)$ signifies the solution of (9.10) subjected to an initial condition (9.11), by a direct computation, where r, t and τ are reflection and transmission operators as discussed in Chapter 6. These operators are well-known and can be obtained as solutions of a one-dimensional Riccati type equation. For example, we have the Cauchy problem,

$$\frac{\partial t}{\partial z}(z; \Omega, \Omega_0) + \frac{1}{v}t(z; \Omega, \Omega_0) \tag{9.19}$$

$$= e^{-z/v_0}\left[p(-\Omega; -\Omega_0) + \frac{1}{4\pi}\int_0^{2\pi} \rho(z; \Omega, \Omega')p(\Omega', -\Omega_0)\frac{d\Omega'}{v'}\right]$$

$$+\frac{1}{4\pi}\int_0^{2\pi} p(-\Omega; -\Omega'')t(z; \Omega'', \Omega_0)\frac{d\Omega''}{v''}$$

$$+\frac{1}{16\pi^2}\int_0^{2\pi}\int_0^{2\pi} \rho(z, \Omega, \Omega')p(\Omega'; -\Omega'')t(z; \Omega'', \Omega_0)\frac{d\Omega'\,d\Omega''}{v'\,v''}$$

with initial condition

$$t(0, \Omega, \Omega_0) \equiv E = \text{identity}. \tag{9.20}$$

Once ρ is known, equation (9.19) is linear and can be easily solved.

Likewise, operators τ and r with initial condition

$$\tau(0, \Omega, \Omega_0) = E \text{ and } r(0, \Omega, \Omega_0) = 0. \tag{9.21}$$

We have accomplished solving a three-dimensional equation (9.10) by using equation (9.18) and solutions of one-dimensional equations. In Chapter 4, we discussed the existence of the inverse $[E - rK]$ which is based on the convergence of the series $E + rK + \cdots + (rK)^n \cdots$. This convergence can be guaranteed by the results of Wang [9], if the system is *conservative*, i.e.,

$$\|p\| = \int_0^{2\pi}\int_0^{2\pi} p(\pm\Omega, \pm\Omega_0)d\Omega d\Omega_0 \le 1 \tag{9.22}$$

and $S(0, x, y; \Omega, \Omega_0)$ is *dissipative*, i.e.,

$$\left\|\int_0^{2\pi}\int_0^{2\pi} S(0, x, y; \Omega, \Omega_0)I(0, x, y; \Omega_0)d\Omega_0 d\Omega\right\| < \left\|\int_0^{2\pi} I(0, x, y; \Omega_0)d\Omega_0\right\| \tag{9.23}$$

for all $I(0, x, y; \Omega)$. In radiative transfer, both conditions can be easily satisfied.

Now the desired asymptotic solution has the form

$$S_n(K) = \rho + tK\left[\sum_{i=0}^{n}(rK)^n\right]\tau \tag{9.24}$$

and

$$\lim_{n\to\infty} S_n(K) = S(K) \tag{9.25}$$

where $S_n(K)$ is a solution with first n-th order of multiple scattering between the atmospheric layer and the ground reflection.

9.5 Approximations

The ground reflection K can be decomposed into two parts,

$$K = K(x, y) = K_g + K_t, \qquad (9.26)$$

where K_g is the ground reflection and K_t is the target reflection , which is restricted to a small area as indicated by equation (9.12). Upon substituting K_g and K_t, respectively, into Equation (9.18), we obtain $R(K_g)$ and $R(K_t)$, see Figures 9.1 and 9.2.

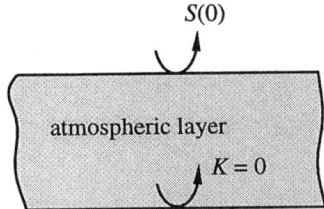

Figure 9.1. Cross section of reflection due to atmospheric layer only.

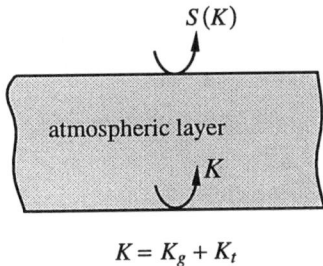

Figure 9.2. Cross section of reflection due to atmospheric layer, target and earth ground.

We also obtain

$$S(K) = -\rho + S(K_g) + S(K_t) + tK_t r K_g \tau + \cdots \qquad (9.27)$$

where $S(K)$ is the total reflection due to the atmospheric layer, target and earth surface. The remaining terms in Equation (9.27) involve higher order interactions between K_g and K_t and the atmospheric layer. The true target reflection K_t can be solved by the formula

$$K_t = [E + Wr]^{-1}W + K_g \qquad (9.28)$$

where

$$W = t^{-1}[S(K) - \rho]\tau^{-1}. \tag{9.29}$$

Equation (9.28) gives the true tangent reflection in terms of $S(K)$, K_g, r, t and τ. In application, $S(K)$ is the observed total reflection above the top of the atmospheric layer, and K_g can be obtained from the observed $S(K_g)$ when the searchlight is off.

As we recall, within a narrow spectrum band, the ground reflection K_g is weak in comparison with the first term in Equation (9.27). Therefore K_g in equation (9.28) can be neglected and it is approximated by

$$K_t \doteq (E + Wr)^{-1}W. \tag{9.30}$$

On the other hand, equation (9.27) reduces to

$$S(K_t) \doteq S(K) - R(K_g) + \rho. \tag{9.31}$$

It is customary to write

$$t = \exp\left(-\frac{1}{v}\int_0^{z_1} \alpha dz\right)\delta(\Omega + \Omega_0) + \tilde{t}(h; \Omega, \Omega_0) \tag{9.32}$$

$$\tau = \exp\left(-\frac{1}{v}\int_0^{z_1} \alpha dz\right)\delta(\Omega + \Omega_0) + \tilde{\tau}(h; \Omega, \Omega_0) \tag{9.33}$$

where z_1 is the optical thickness of the atmospheric layer and $\alpha = \alpha(z)$ is the attenuation coefficient. In both equations, the first terms are the specular part and the second terms are the diffuse parts.

Substitute equations (9.32) and (9.33) into equation (9.31) and consider the first order of multiple scattering between the atmospheric layer and the ground reflection, to obtain the first order approximation solution,

$$K_t^{(1)} = [S(K_t) - \rho]\exp\left[-\frac{2}{v}\int_0^{z_1} \alpha dz\right]\delta(\Omega + \Omega_0). \tag{9.34}$$

By equation (9.31),

$$K_t^{(1)} \doteq [S(K) - S(K_g)]\exp\left[-\frac{2}{v}\int_0^{z_1} \alpha dz\right]\delta(\Omega + \Omega_0). \tag{9.35}$$

This equation gives the first order approximation to the true target reflection in terms of $S(K)$ and $S(K_g)$. As we recall $S(K)$ and $S(Kg)$, in practice, are observed intensities.

Likewise, we can obtain the second order approximation of K_t, that is,

$$K_t^{(2)} = [S(K) - S(K_g)][E + rK_g]^{-1}\exp\left[-\frac{2}{v}\int_0^{z_1} \alpha dz\right] \cdot \delta(\Omega + \Omega_0). \tag{9.36}$$

Higher order approximations can be constructed by this method, but the results are more complicated for computation and may be unnecessary when $S(K_t) \gg S(K_g)$.

9.6 Numerical Simulation

A numerical simulation for the searchlight problem without background light has been carried out under somewhat restricted initial and boundary conditions via the Monte Carlo method. Assume that the atmosphere under consideration is optically uniform throughout, bounded with horizontally uniform Lambertian reflection. Furthermore, the searchlight is perpendicularly incident on the origin of Cartesian coordinates.

Under such initial boundary conditions the numerical simulation of multiple scattering processes of photons is carried out by the Monte Carlo method via Equation (9.2). The free path of photons (l) is calculated using a uniformly distributed random number r_u,

$$l = C_r \frac{\tau}{z_1} \log r_u \tag{9.37}$$

where C_r is a parameter which regulates the path length after reflection. In the case of reflection on a flat surface the value of C_r is equal to 1.42, whereas C_r is unity after re-emission by atmospheric particles.

Several simulations were performed and in each case the optical thickness of the atmosphere had a different value between 0.01 and 10.0. It is assumed that the scattering coefficient of atmospheric particles (λ) is 1.0; the ground albedo A is 0.4; the incident direction of the photons is normal, i.e., $(\theta_0, \phi_0) = (0, 0)$; the total number of photons is 3.2 million; and the aperture of the photosensor at the top of the atmospheric layer is 0.2.

Then the radiation intensity under consideration for incident flux F at the top is estimated as below.

$$I(z_1, 0, 0; \Omega) = \pi F \frac{\Delta N(z_1, 0, 0; \Omega)}{v N \Delta s}, \tag{9.38}$$

where Δs is the surface area of the aperture at the top of the atmospheric layer and ΔN is the number of photons passing through the aperture in direction Ω. The number of incident photons from the searchlight N is conserved throughout the whole process.

Under such initial and boundary conditions the polar dependence of the reflected searchlight intensity at the origin of the top of the atmospheric layer $I(z_1, 0, 0; \Omega)$ is numerically simulated in Figure 9.3 and Figure 9.4. It shows that the intensity contours at the origin of the top of the atmospheric layer for normal incidence have a similar tendency. In other words, from the horizontal direction to the normal, the radiation intensity increases monotonically, arriving at a maximum for normal incidence. As the optical thickness increases, the sharp peak near $v = 1(\theta_0 = 0)$ disappears.

The statistical error in the simulation depends on the number of emitted photons, the optical thickness, the polar angle of reflected photon and the differential coefficient of estimated curves. It can be expected to be in the range between one and ten percent in our calculation.

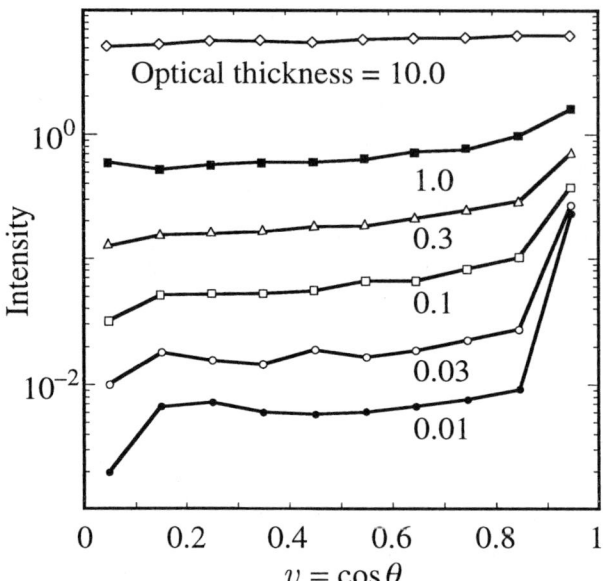

Figure 9.3. Polar angle dependence of reflected intensity. Case 1: Searchlight on the top of atmosphere.

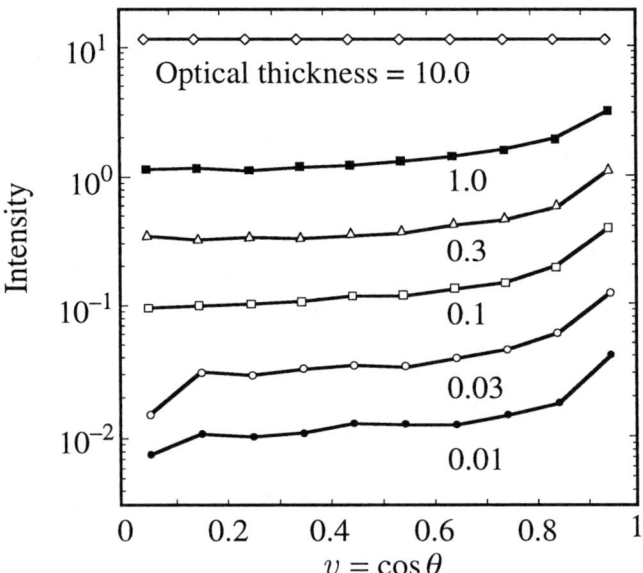

Figure 9.4. Polar angle dependence of reflected intensity. Case 2: Searchlight on the bottom of the atmosphere.

9.7 Conclusions

In this chapter we have constructed a mathematical model for the searchlight problem. Similar to previous chapters, the method of construction is invariant and the result is an initial valued differential equation. The computational method discussed in Chapters 2 and 3 may be used here. We also presented approximation solutions in general and the results of a numerical simulation of a searchlight perpendicularly incident on the atmosphere with the earth's surface being a Lambertian reflector.

References

1. S. Ueno, "Invariant Imbedding and Diffuse Reflection of Radiation by Multidimensional Slab," *Memoires of Kanuzawa Institute of Technology*, A, No. 6, 1976.

2. S. Chandrasekhar, *Radiative Transfer*, Oxford University, New York, 1950.

3. J. Lenoble, "Solution of Radiative Transfer in an Infinite Scattering Medium Illuminated by a Point Source," Scientific Report No. 6, Department of Meteorology, *University of California*, 1960.

4. R. W. Preisendorfer, "The Fundamental Method of Solving Point Source Problems in Discrete-Space Radiative Transfer Theory," SIO reference 69-71, *Scripps Institute of Oceanography*, 1959.

5. R. Bellman, R. Kalaba and S. Ueno, "Invariant Imbedding and Diffuse Reflection of Radiation from a Collimated Point Source I and II," *Rand Corporation*, RM-3141, ARPA, 1962.

6. R. Bellman, R. Kalaba and S. Ueno, "Invariant Imbedding and Diffuse Reflection of Radiation from a Collimated Point Source III," *Rand Corporation*, RM-3166, ARPA, 1962.

7. A. P. Wang and S. Ueno, "Searchlight on a Target with Diffuse Background I," *J. Math. Physics*, Vol. 34, No. 2, February 1993, pp. 878–884.

8. A. P. Wang and S. Ueno, "Searchlight on a Target with Diffuse Background II," *Computers Math. Applic.*, Vol. 27, No. 9/10, 1994, pp. 169–172.

9. A. P. Wang, "The Four-Ports Dynamic System," *J. of Math. Analysis and Applications*, Vol. 120, No. 2, 1986, pp. 416–437.

10. A. P. Wang and S. Ueno, "Searchlight on a Target with Diffuse Background I," *J. of Mathematical Physics*, Vol. 34, 1993, pp. 878–884.

10. Transfer of Radiation with Spherical Symmetry

This chapter on radiative transfer in a spherical shell medium is in two parts: first, the construction of linear-operator equations and their reduction to a class of functional equations; then, the description of numerical techniques for dealing with the functional equations and the presentation of computational results. These analytical and computational results are applicable to terrestrial and stellar atmospheres. In the analytical theory, we treat with inhomogeneous anisotropically scattering shells with internal or external illumination, and with reflecting or absorbing cores. The computational results presented herein are for homogeneous shells, and can be extended to inhomogeneous and anisotropically scattering ones.

10.1 Introduction

The analysis of radiative transfer in plane-parallel media is extended to take curved surfaces into consideration. This has led to the study of radiative transfer in spherical symmetry by Bellman, Kagiwada, Ueno, Wang, etc., [1–12]. The analysis presented here is based on invariant imbedding and the scattering matrix, as discussed in Chapters 2 and 6.

By a proper interpretation of the meaning of intensities and linear operators, we construct a system of rather general linear-operator equations which govern the radiation field in spherical symmetry with external illumination and with internal illumination. We have succeeded in solving Chandrasekhar's problem in a spherical shell, Schuster's problem in the theory of line formation and the Milne problem of the diffusion of light from a central star, as special cases.

10.2 Intensity and Operations

The medium in which the transfer of radiation takes place has spherical symmetry, and the medium's properties are functions of the depth from the center of the sphere. Incident radiation is spherically uniform. The medium may be an inhomogeneous spherical shell of atmosphere which scatters anisotropically, with radii x and y, $0 < x \leq y$. The intensity in the total radiation field at radius z, $x \leq z \leq y$, at inclination $\cos^{-1} u$, $0 < u \leq 1$, to the radius vector directed in the inward direction is denoted by $I(z, -u)$ and the intensity in the outward direction with the inclination $\cos^{-1} u$, $0 < u \leq 1$, to the radius vector is $I(z, u)$.

Let the total radiation field in the inward direction at the outer surface of the spherical shell be $I(y, -u)$, $0 < u \leq 1$. Then the total intensities transmitted and reflected are denoted by $I(x, -v)$ and $I(y, v)$, respectively. The transmitted radiation is composed of two parts,

$$I(x, -v) = I^s(x, -v) + I^d(x, -v), \tag{10.1}$$

the specular part (the directly transmitted part) and the diffuse part. However, the reflected radiation field has the diffuse part only,

$$I(y, v) = I^d(y, v). \tag{10.2}$$

Viewing $I^s(x, -v)$, $I^d(x, -v)$ and $I^d(y, v)$ as outputs of a linear system due to an input $I(y, -u)$, we may write

$$I^s(x, -v) = \mathbf{Q} \cdot I(y, -u), \quad I^d(x, -v) = \tau \cdot I(y, -u),$$

$$I^d(y, v) = \rho \cdot I(y, -u), \tag{10.3}$$

where $\mathbf{Q}$, τ and ρ are called the *specular, transmission,* and *reflection operators*. The operators have integral representations with kernels $Q(x, y, -v, -u)$, $\tau(x, y, -v, -u)$ and $\rho(x, y, v, -u)$, respectively. For example,

$$[\tau \cdot I(y, -u)](x, -v) = \int_0^1 \tau(x, y, -v, -u) I(y, -u) du. \tag{10.4}$$

Similarly, if there is an incident radiation field $I(x, u)$ at x in the outward direction, then the corresponding operators are denoted by $\mathbf{P}$, t, r and their kernels are $P(x, y, v, u)$, $t(x, y, v, u)$, $r(x, y - v, u)$, respectively. For more details on representations of these linear operators, see [4].

It should be noted that the intensities we define here have taken all multiple scattering into account. With the exception that the core is a perfect absorber, there is multiple scattering taking place between the core and the spherical shell. This includes the case when the core is a vacuum. For example, if the core radiates an amount I_i of intensity at the surface in the outward direction of the intensity $I(x, v) = (\mathbf{E} - \mathbf{K} \cdot \mathbf{r})^{-1} I_i$, where $\mathbf{K}$ is the reflection operator for the core, (see Section 10.6). There are two basic physical distinctions in the radiative transfer in slab geometry versus the spherical geometry. The first one is that when a slab is imbedded in a vacuum or a nonreflective space, the intensity at once leaves the slab and will not return. Therefore there are no multiple reflections taking place between the slab and the vacuum space. However, this is not true if a spherical shell is imbedded in a vacuum with a nonreflective core. There is multiple scattering taking place between the core and the spherical shell. Another distinction is that while a ray of radiation travels in a straight line in a spherical shell, its parameter u changes with radius; see Fig. 10.1. In the slab geometry it remains the same.

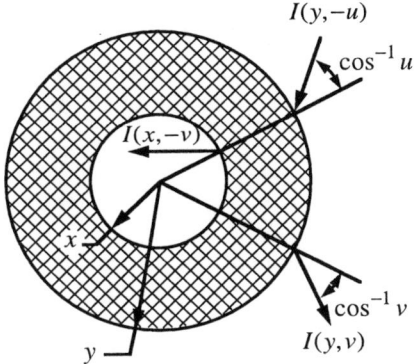

Figure 10.1. Spherical symmetrical shell with incident radiation field $I_i = I(y, -u)$ at the outer surface.

10.3 Transfer of Radiation

Let us consider the intensities at radii $z - \Delta z$ and z, see Fig. 10.2.

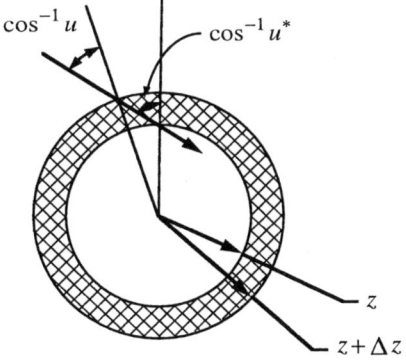

Figure 10.2. Changing of $\cos^{-1} u$ due to a small change of radius, where $u^* = \{1 - [(z + \Delta z)/z]^2 (1 - u)^2\}^{1/2}$.

Then the change in the intensity of the radiation field with respect to z due to moving from one radius to another involves three parts. The first and second parts are the volume attenuation and the diffusion parts,

$$\frac{\alpha(z)}{v}I(z,v) - \frac{\sigma(z)}{2v}\int_{-1}^{1}p(z,v,u)I(z,u)du, \quad 0 < |v| \le 1, \tag{10.5}$$

as in the slab case, where $\alpha(z)$, $\sigma(z)$ and $p(z,v,u)$ are, respectively, the volume attenuation, scattering coefficient, and azimuth independent phase function. Intensity is a function of z and v. The change in cosine of the angle due to the change of radius constitutes the third part,

$$-\frac{1-v^2}{zv}\frac{\partial}{\partial v}I(z,v). \tag{10.6}$$

The total effect is given by superposition. For the moment, let us consider the specular part of the radiation only, i.e., the diffuse part of the contribution is ignored. The inward specular equation is given by

$$\frac{\partial}{\partial z}I^s(z,-v) = \left(\frac{\alpha(z)}{v} - \frac{1-v^2}{zv}\frac{\partial}{\partial v}\right)I^s(z,-v), \quad x \le z \le y, \quad 0 < v \le 1. \tag{10.7}$$

This partial differential equation has a general solution of the form

$$\Psi[z(1-v^2)^{1/2}],$$

where Ψ is a differentiable function in z and v. In particular, if the boundary condition at y in the inward direction is given by $\delta(v-u)$, then a particular solution for (10.7) is

$$I^s(z,-v) = \delta(v_* - u)e^{-h(z,y,u)}, \tag{10.8}$$

where the optical thickness

$$h(z,y,u) = \int_z^y \frac{\alpha(\xi)}{u^*(\xi)}d\xi$$

and the superscript * and subscript * are used to denote the following relationships:

$$f_* = f_*(z,x) = \left[1 - \left(\frac{z}{x}\right)^2(1-f^2)\right]^{1/2}$$

$$f^* = f^*(z,x) = \left[1 - \left(\frac{x}{z}\right)^2(1-f^2)\right]^{1/2}, \qquad z < x.$$

We see that $(f_*)^* = (f^*)_* = f$.

Upon substituting the Q defined in (10.3) into eq. (10.7), we have

$$\frac{\partial}{\partial z}\mathbf{Q} = \left(\frac{\alpha(z)}{v} - \frac{1-v^2}{zv}\frac{\partial}{\partial v}\right)\mathbf{Q}, \tag{10.9}$$

where the operator $\mathbf{Q}$ has kernel $Q(z,y,-v,-u)$. It is clear that

$$Q(z, y, -v, -u) = \delta(v_* - u)e^{-h(z,y,u)} \tag{10.10}$$

is a solution of eq. (10.9) with initial value $\delta(v - u)$ as $z = y$. Under the integral convention, see eq. (10.4), $\mathbf{Q}$ acts as an identity when $z = y$ which agrees with the physics of the system.

We have chosen a simple case to show the derivation of the intensity and the operator equations. The solution for the intensity equation depends on the value of the incident radiation while the operator equation is independent of the incident radiation field. Also it should be noted that the Dirac delta δ used in eq. (10.10) should be considered as an operator in the sense of distribution theory. More precisely, in the integral representation in eq. (10.4), all operators and the Dirac delta are considered as in the sense of regular distribution, while the Dirac delta used in (10.8) is not a well-defined function.

Let us compute the total flux at z, $x \le z \le y$ with $\alpha(z) \equiv 0$ in eq. (10.9) and $u_c \le u \le 1$, with $u_c^2 = 1 - (x/y)^2$. Then

$$4\pi z^2 \int_0^1 \int_0^1 v_* Q(z, y, -v, -u)I^s(y, -u)dvdu$$

$$= 4\pi z^2 \int_u^1 \int_0^1 v_* \delta(v_* - u)I^s(y, -u)\frac{dv}{dv_*}dv_*du$$

$$= 2\pi x^2 \int_0^1 uI^s(y, -u)du.$$

That is, in the case of $\alpha(z) \equiv 0$ and $u_c \le u \le 1$, the total specular flux is independent of z.

For $0 < u < u_c$, the specular part of the intensity will not reach the area in the spherical shell with radius less than $y_c = y(1 - u^2)^{1/2}$. However, this part of the intensity is passing through the spherical shell and it becomes a reflected part (see Fig. 10.3). In fact,

$$I^s(y, v) = \delta(v - u)e^{-2h(y_c, y, u)}, \quad 0 < u < u_c.$$

This situation does not appear in the slab case, since $u_c = 0$. And this situation also holds for a very thin spherical shell for a somewhat different reason. For incident intensity with inclination $\cos^{-1} u$, $0 < u < 1$, we can always choose Δz so small that

$$u \ge u_c = \{1 - [(z - \Delta z)/z^2]\}^{1/2}.$$

In a similar manner, the specular operator equation for $\mathbf{P}$ is

$$\frac{\partial}{\partial z}\mathbf{P} = -\left(\frac{\alpha(z)}{v} + \frac{1 - v^2}{zv}\frac{\partial}{\partial v}\right)\mathbf{P} \tag{10.11}$$

with initial value

$$\mathbf{P} = \mathbf{E} = \text{ identity for } z = y.$$

Here the operator $\mathbf{P}$ has kernel $P(y, z, v, u)$, and

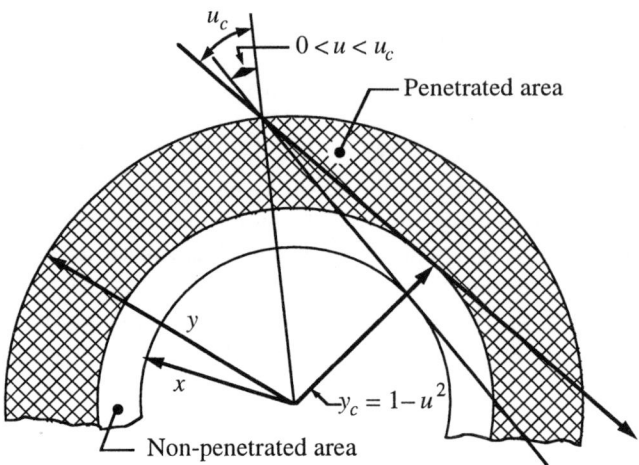

Figure 10.3. Section of spherical shell with specular part of radiation only, showing nonpenetrated area.

$$P(y, z, v^*, u) = \delta(v^* - u)e^{-h(y,z,u)} \tag{10.12}$$

is a solution. We shall not discuss the properties of P, since they are analogs of the properties of Q.

10.4 Coefficients of the Medium

The differential operator-equations for the radiative transfer may be obtained when the added spherical shell is very thin. The medium coefficients at y are specified by behavior of such a thin shell extended from y to $y + \Delta y$ where Δy is small. The diffusion coefficients $\mathbf{a}^*$, $\mathbf{b}^*$, $\mathbf{c}^*$, and $\mathbf{d}^*$ are defined as the limits of diffusion operators ρ, $\mathbf{t}$, $\mathbf{r}$ and τ associated with the thin shell. Kernels of $\mathbf{a}^*$, $\mathbf{b}^*$, $\mathbf{c}^*$, $\mathbf{d}^*$ can be represented by phase functions, which we have already used in the previous section. Phase functions are physical parameters and are also called differential cross section. In the operator notation we write

$$\begin{pmatrix} \mathbf{b}^* & \mathbf{a}^* \\ \mathbf{c}^* & \mathbf{d}^* \end{pmatrix} = \frac{\sigma}{2v} \begin{pmatrix} \mathbf{p}^{++} & \mathbf{p}^{+-} \\ \mathbf{p}^{-+} & \mathbf{p}^{++} \end{pmatrix}, \tag{10.13}$$

where the operator $p^{\pm\pm}$ has kernel $p(y, \pm v, \pm u)$.

To obtain the specular coefficients, we shall consider the case operators $\mathbf{Q}$ and $\mathbf{P}$ to be associated with the same thin spherical shell. Since $\mathbf{Q}$ and $\mathbf{P}$ involve $\delta(v_* - u)$ and $\delta(v^* - u)$, the limits of a left-hand and right-hand specular operators under the composite operation are somewhat different. For this reason, let us compute two typical cases.

In the first case, we consider the limit of $\mathbf{P}$ on the left-hand side of a composite operation $\mathbf{P}$. That is,

$$\lim_{\Delta y \to 0} \frac{1}{\Delta y}[\mathbf{P} \cdot \rho - \rho](y, v) \tag{10.14}$$

$$= \lim_{\Delta y \to 0} \frac{1}{\Delta y} \int_0^1 [\delta(v^* - w)e^{h(y,y+\Delta y,w)} - \delta(v - w)]\rho(x, y, w, -u)\,dw$$

$$= -\left(\frac{\alpha(y)}{v} + \frac{1 - v^2}{yv}\frac{\partial}{\partial v}\right)\rho(x, y, v, -u),$$

where the last equality follows from eq. (10.11) and $P(y, y + \Delta y, -v, -w) = \delta(v - w)$ when $\Delta y \to 0$.

In the second case we consider the limit of $\mathbf{Q}$ on the right-hand side of a composite operation $\rho \cdot \mathbf{Q}$. Let us compute

$$[\rho \cdot \mathbf{Q}](y, v) = \int_0^1 \rho(x, y, v, -w)\delta(w_* - u)e^{-h(y,y+\Delta y,u)}\,dw$$

$$= \left(\frac{y + \Delta y}{y}\right)^2 \left(\frac{u}{u^*}\right) \int_0^{u_c} \rho(x, y, v, -w)\delta(w - u^*)\,dw,$$

where $u^* = u^*(y, y + \Delta y)$ and $u_c = \{1 - [(y + \Delta y)/y]\}^{1/2}$. From this it follows that, by taking the limit,

$$\lim_{\Delta y \to 0} \frac{1}{\Delta y}[\rho \cdot \mathbf{Q} - \rho](y, v)$$

$$= -\int_0^{u_c} \rho(x, y, v, -w)$$

$$\cdot \left(\frac{\alpha(y)}{w} - \frac{2}{y} - \frac{1 - w^2}{yw^2} + \frac{1 - w^2}{yw}\frac{\partial}{\partial w}\right)\delta(w - u^*)\,dw|_{u^*=u}$$

$$= -\left(\frac{\alpha(y)}{u} - \frac{2}{y} - \frac{1 - u^2}{u^2 y} - \frac{1 - u^2}{yu}\frac{\partial}{\partial u}\right)\rho(x, y, v, -u), \tag{10.15}$$

where we used eq. (10.9) and the fact that, for $0 < u \le 1$,

$$0 = \frac{\partial}{\partial w}\int_0^{u_c} \rho(x, y, v, -w)\delta(w - u^*)\,dw$$

$$= \rho(x, y, v, -w)\delta(w - u^*)|_{w=0}^{w=u_c}$$

since

$$u^* = u^*(y, y + \Delta y) = \left[1 - \left(\frac{y + \Delta y}{y}\right)^2(1 - u^2)\right]^{1/2} \ne u_c$$

or 0. In view of (10.14) and (10.15) where ρ is arbitrary, we have that the specular coefficients at y corresponding to **P** and **Q** are coefficients **B** and $\widetilde{\mathbf{D}}$, and their kernels have the form

$$B(y, v, u) \quad = -\left(\frac{\alpha(y)}{v} + \frac{1-v^2}{yv}\frac{\partial}{\partial u}\right)\delta(v - u),$$

$$\widetilde{D}(y, -v, -u) \quad = -\left(\frac{\alpha(y)}{v} - \frac{2}{y} - \frac{1-u^2}{yu}\frac{\partial}{\partial u}\right)\delta(v - u),$$

(10.16)

with the understanding that $B(y, v, u)$ is used in the left-hand of a composite operation and $\widetilde{D}(y, -v, -u)$ in the right-hand, signified by superscript $\sim$. In a similar manner, the operators $\widetilde{B}$ and D have kernels

$$\widetilde{B}(y, v, u) \quad = -\left(\frac{\alpha(y)}{u} + \frac{2}{y} + \frac{1-u^2}{yu^2} - \frac{1-u^2}{yu}\frac{\partial}{\partial u}\right)\delta(v - u),$$

$$D(y, -v, -u) \quad = -\left(\frac{\alpha(y)}{u} + \frac{1-v^2}{yu}\frac{\partial}{\partial v}\right)\delta(v - u).$$

(10.17)

By (10.16) and (10.17), eqs. (10.9) and (10.12) can be expressed as

$$\frac{\partial}{\partial y}\mathbf{Q} = \mathbf{Q}\cdot\widetilde{\mathbf{D}} \quad \text{and} \quad \frac{\partial}{\partial y}\mathbf{P} = \mathbf{B}\cdot\mathbf{P}.$$

(10.18)

10.5 State and Local Form

With coefficients as constructed in eqs. (10.13), (10.16), and (10.17), one can obtain a complete system of equations which govern the radiative transfer under consideration. This is done by the invariant imbedding method of adding a thin layer and using the technique of star-products. The method was developed originally for the slab. Mathematically, the theory is constructed on the class of linear operators on Hilbert space. Therefore, it can apply equally well to the spherical shell provided the meaning of the intensity is understood in the way defined here; see remarks at the end of the second section. We take the star-product of the scattering matrix associated with radii x and y, $0 < x \leq y$, defined by

$$S = \begin{pmatrix} \mathbf{P} & \mathbf{0} \\ \mathbf{0} & \mathbf{Q} \end{pmatrix} + \begin{pmatrix} \mathbf{t} & \rho \\ \mathbf{r} & \tau \end{pmatrix},$$

(10.19)

with another scattering matrix associated with radii y and $y + \Delta y$, then take the limit as $\Delta y \to 0$. With the aid of eqs. (10.13), (10.16), (10.17), and (10.18), we obtain a set of operator equations

$$t_y = (\mathbf{b} + \rho\cdot\mathbf{c})(\mathbf{t} + \mathbf{P}) - \mathbf{B}\cdot\mathbf{P},$$

(10.20a)

$$\rho_y = \mathbf{a} + \mathbf{b} \cdot \rho + \rho \cdot \mathbf{d} + \rho \cdot \mathbf{c} \cdot \rho, \tag{10.20b}$$

$$r_y = (\tau + \mathbf{Q}) \cdot \mathbf{c} \cdot (\mathbf{t} + \mathbf{P}), \tag{10.20c}$$

$$\tau_y = (\tau + \mathbf{Q}) \cdot (\mathbf{d} + \mathbf{c} \cdot \rho) - \mathbf{Q} \cdot \tilde{\mathbf{D}}, \tag{10.20d}$$

where the subscript y denotes partial differentiation, and where the coefficients

$$\begin{pmatrix} \mathbf{b} & \mathbf{a} \\ \mathbf{c} & \mathbf{d} \end{pmatrix} = \begin{pmatrix} \mathbf{B} & \mathbf{0} \\ \mathbf{0} & \mathbf{D} \end{pmatrix} + \begin{pmatrix} \mathbf{b}^* & \mathbf{a}^* \\ \mathbf{c}^* & \mathbf{d}^* \end{pmatrix} \tag{10.21}$$

are evaluated at y.

In the slab case, one will obtain another set of operator-equations by adding a thin layer at x. There is a remarkable symmetry between the partial differentiation with respect to y and that with respect to x. However, for the spherical shell one should not expect complete symmetry, since from the physical point of view the shell being added at y is different from that added at x. This also can be explained from the mathematical point of view: under the composite operation the left-hand coefficient is different from the right-hand coefficient, as discussed in the previous section. However, the theory of an additional layer and star-products is still valid. Details are not presented here; we merely state the result:

$$-\mathbf{t}_x = \mathbf{t} \cdot (\bar{\mathbf{b}} + \mathbf{a} \cdot \mathbf{r}) - \mathbf{P} \cdot \tilde{\mathbf{B}}, \tag{10.22a}$$

$$-\rho_x = (\mathbf{t} + \mathbf{P}) \cdot \mathbf{a} \cdot (\tau + \mathbf{Q}), \tag{10.22b}$$

$$-\mathbf{r}_x = \mathbf{c} + \bar{\mathbf{d}} \cdot \mathbf{r} + \mathbf{r} \cdot \bar{\mathbf{b}} + \mathbf{r} \cdot \mathbf{a} \cdot \mathbf{r}, \tag{10.22c}$$

$$-\tau_x = (\bar{\mathbf{d}} + \mathbf{r} \cdot \mathbf{a}) \cdot \tau - \mathbf{D} \cdot \mathbf{Q}, \tag{10.22d}$$

with $\bar{\mathbf{b}} = \tilde{\mathbf{B}} + \mathbf{b}^*$, $\bar{\mathbf{d}} = \mathbf{D} + \mathbf{d}^*$ and the coefficients evaluated at x. Eqs. (10.20) and (10.22) are called the *state forms for a symmetric spherical shell*. They correspond to the state form for a slab if $u^* = u_* = u$ and $v^* = v_* = v$. In this case x and y are slab depth.

Let us consider the intensities on both sides of a thin spherical shell with radii z and $z + \Delta z$ where Δz is small. We have

$$I(z + \Delta z, v) = (\mathbf{P} + \mathbf{t}) \cdot I(z, u) + \rho \cdot I(z + \Delta z, -u),$$

$$I(z, -v) = \mathbf{r} \cdot I(z, u) + (\mathbf{Q} + \tau) \cdot I(z + \Delta z, -u),$$

where operators are associated with the thin spherical shell. Upon taking the limits as $\Delta z \to 0$ and using coefficients as stated in eqs. (10.13), (10.16), and (10.17), the following linear system is obtained:

$$\frac{\partial}{\partial z}\begin{pmatrix} I(z,v) \\ I(z,-v) \end{pmatrix} = \left[\begin{pmatrix} \mathbf{B} & \mathbf{0} \\ \mathbf{0} & -\mathbf{D} \end{pmatrix} + \begin{pmatrix} \mathbf{b}^* & \mathbf{a}^* \\ -\mathbf{c}^* & -\mathbf{d}^* \end{pmatrix} \right] \cdot \begin{pmatrix} I(z,v) \\ I(z,-v) \end{pmatrix}.$$
$$\tag{10.23}$$

This intensity-equation is called the *local form* for a symmetric spherical shell and all coefficients are evaluated at z, $x \le z \le y$. It is presented in the decomposed form, the first term on the right-hand side of eq. (10.23) being the specular part and the second term being the diffuse part. Using eq. (10.21), the above local form appears identical to that for a slab, with somewhat different meanings for the coefficients.

10.6 The Reflecting Core

Let us consider a spherical symmetric shell of atmosphere surrounding a reflecting core with a reflection governed by an operator $\mathbf{K}$. We consider two types of problems. In problem type A the shell is externally illuminated, while in problem type B the shell is internally illuminated. See Figs. 10.4 and 10.5, where I_i is the illuminating radiation. These are the fundamental problems in the theories of the illumination of the sky and of the planetary illumination.

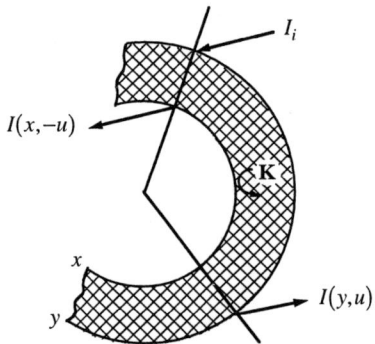

Figure 10.4. A section of a spherical shell with external illumination I_i.

The basic tasks for problems of type A and B are the determination of the intensity field at the outside of the spherical shell in the outward direction, $I(y, u)$, and the specification of the radiation field, $I(x, -u)$, as seen by an observer at the surface of the core looking at the atmosphere.

Before we proceed with the physical meanings of such problems, we shall give a discussion of the operator $\mathbf{K}$. The precise mathematical meaning of $\mathbf{K}$ is that, when there is no spherical shell, it relates $I(x, u)$ and $I(x, -v)$ by

$$I(x,u) = \mathbf{K} \cdot I(x,-v). \tag{10.24}$$

When the core is a vacuum, then $\mathbf{K} = \mathbf{V}$ where $\mathbf{V}$ has kernel $\delta(v + u)$ since there is no absorption or diffusion taking place, and all incident intensities

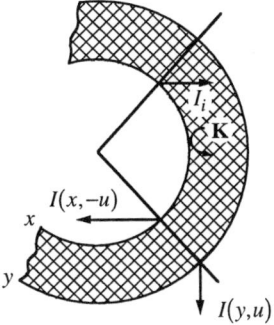

Figure 10.5. A section of spherical shell with internal illumination I_i.

$I(x, u)$ go through the core and constitute an output $I(x, -u)$ where the total flux is preserved (see Section 2), but the direction is reversed. When a slab is imbedded in a vacuum, since there is no intensity returning from the vacuum, we have $\mathbf{K} = \mathbf{0} =$ zero operator. Another case in a spherical shell, $\mathbf{K} = \mathbf{V}$, occurs when the core is made of a perfect reflecting material. On the other hand, if the core is made of a perfect absorbing material, such as a perfect black body, then $\mathbf{K} = \mathbf{0}$. Of course, similar situations hold if a slab is imbedded in a perfectly reflecting or absorbing material.

Problem type A, when $\mathbf{K} = \mathbf{0}$ corresponds to the "standard radiative transfer" with spherical symmetry. When $\mathbf{K}$ has the kernel $K = K(v, u) = Avu$, where A is a constant, we have "Chandrasekhar's problem" with spherical symmetry. When $\mathbf{K} = \mathbf{V}$, we have the problem of a spherical shell imbedded in a vacuum. We may view problem type A as a sort of generalized Chandrasekhar's problem in a spherically symmetric shell.

In problem type B, when $\mathbf{K} = 0$, we have "Schuster's problem" in the theory of line formation by a perfectly absorbing core. When $\mathbf{K} = Avu$ we have the problem of a Lambert reflecting core, which radiates intensities into the surrounding spherical shell of atmosphere. When $\mathbf{K} = \mathbf{V}$, we have "Milne's problem" of the diffusion of light from a star.

We shall attack those two types of problems by constructing a single ideal model and take advantage of the star-product.

Instead of considering a core with given reflection operator K, we assume that there is another spherical shell inside of the given one with radii x_1 and x, $0 < x_1 \leq x$. This inner shell has scattering matrix

$$S_1 = \begin{pmatrix} t_1 & \mathbf{K} \\ r_1 & \tau_1 \end{pmatrix},$$

where t_1 is chosen so that $t_1 \cdot I(x_1, v) = I_i$ and τ_1 and r_1 are unrestricted. The condition imposed on t_1 is required only for problem type B). The replacing of a core by such an inner spherical shell does not alter the results of our problems.

Using the star-product of S_1 and S, see Chapter 4, and $(E - r \cdot K)$ nonsingular, we have

$$
S_1 * S = \begin{pmatrix} (\mathbf{P} + \mathbf{t}) \cdot (\mathbf{E} - \mathbf{K} \cdot \mathbf{r})^{-1} t - 1 \\ \mathbf{r}_1 + \tau_1 \cdot \mathbf{r} \cdot (\mathbf{E} - \mathbf{K} \cdot \mathbf{r})^{-1} \cdot t_1 \\ \rho + (\mathbf{P} + \mathbf{t}) \cdot \mathbf{K} \cdot (\mathbf{E} - \mathbf{r} \cdot \mathbf{K})^{-1} \cdot (\mathbf{Q} + \tau) \\ \tau_1 \cdot (\mathbf{E} - \mathbf{r} \cdot \mathbf{K})^{-1} \cdot (\mathbf{Q} + \tau) \end{pmatrix}
\qquad (10.25)
$$

Eq. (10.25) denotes the overall transmission and reflection by taking account of multiple scattering. The overall reflection at radius y is

$$
\rho(K) = \rho + (\mathbf{P} + \mathbf{t}) \cdot \mathbf{K} \cdot (\mathbf{E} - \mathbf{r} \cdot \mathbf{K})^{-1} \cdot (\mathbf{Q} + \tau), \qquad (10.26)
$$

where t, τ, ρ and r are solutions of (10.20) and (10.22) subjected to the initial conditions

$$
S = \begin{pmatrix} \mathbf{E} & 0 \\ 0 & \mathbf{E} \end{pmatrix}, \quad \text{when } x = y.
$$

The reflected radiation at y in the outward direction due to the incident radiation I_i at y is

$$
I(y, v) = I^*(y, v) = \rho(K) \cdot I_i. \qquad (10.27)
$$

This result is the desired first part of the answer for problems of type A. To obtain the remaining part of the result, we note that the overall transmission at x_1 due to the incident radiation I_i at y is

$$
\tau_1 \cdot (\mathbf{E} - \mathbf{r} \cdot \mathbf{K})^{-1} \cdot (\mathbf{Q} + \tau) \cdot I_i.
$$

The total transmitted intensity at the surface of the core in the inward direction is given by

$$
I(x, -v) = (\mathbf{E} - \mathbf{r} \cdot \mathbf{K})^{-1} \cdot (\mathbf{Q} + \tau) \cdot I_i. \qquad (10.28)
$$

Eq. (10.28) should be separated into two parts, the specular part and the diffuse part, by

$$
\begin{aligned}
I^s(x, -v) &= \mathbf{Q} \cdot I_i, \\
I^d(x, -v) &= \tau(K) \cdot I_i,
\end{aligned}
\qquad (10.29)
$$

where

$$
\tau(K) = (\mathbf{E} - \mathbf{r} \cdot \mathbf{K})^{-1} \cdot [\tau + \mathbf{r} \cdot \mathbf{K} \cdot \mathbf{Q}]. \qquad (10.30)
$$

Thus the problem of type A is solved for arbitrary K provided $(\mathbf{E} - \mathbf{r} \cdot \mathbf{K})$ is nonsingular.

For the problem of type B, we use the fact that the intensities at x can be expressed as

$$
I(x, -v) = \mathbf{r} \cdot I(x, u)
$$

and

$$
I(x, v) = \mathbf{t}_1 \cdot I(x_1, u) + \mathbf{K} \cdot I(y, -u) = I_i + \mathbf{K} \cdot I(y, -u).
$$

The last equality follows from the construction of $\mathbf{S}_1$. By eliminating $I(x, -v)$, we have

$$I(x, v) = (\mathbf{E} - \mathbf{K} \cdot \mathbf{r})^{-1} \cdot I_i. \tag{10.31}$$

It follows immediately that

$$I(y, v) = (\mathbf{P} + \mathbf{t}) \cdot I(x, u) = (\mathbf{P} + \mathbf{t}) \cdot (\mathbf{E} - \mathbf{K} \cdot \mathbf{r})^{-1} \cdot I_i$$

since there is no incident radiation at y in the inward direction. Therefore, the specular and diffuse parts at y in the outward direction are

$$\begin{aligned} I^s(y, v) &= \mathbf{P} \cdot I_i, \\ I^d(y, v) &= \mathbf{t}(K) \cdot I_i, \end{aligned} \tag{10.32}$$

where

$$\mathbf{t}(K) = [\mathbf{t} + \mathbf{P} \cdot \mathbf{K} \cdot \mathbf{r}] \cdot (\mathbf{E} - \mathbf{K} \cdot \mathbf{r})^{-1}. \tag{10.33}$$

Eqs. (10.32) are the transmitted intensities at y due to an incident radiation I_i at the surface of the core in the outward direction. The reflected intensity at x is

$$I(x, -v) = I^d(x, -v) = \mathbf{r} \cdot I(x, u) = \mathbf{r}(K) \cdot I_i, \tag{10.34}$$

where

$$\begin{aligned} \mathbf{r}(K) &= \mathbf{r} \cdot (\mathbf{E} - \mathbf{K} \cdot \mathbf{r})^{-1} \\ &= r + r \cdot k \cdot r + r \cdot K \cdot r \cdot Kr + \cdots. \end{aligned} \tag{10.35}$$

The initial values are $\rho(K) = \mathbf{K}$ and zero for $\mathbf{t}(K)$, $\tau(K)$ and $\mathbf{r}(K)$ at $x = y$. This statement can be easily checked by taking the limit as $x \to y$. It also agrees with the physics of our system.

Eqs. (10.32)–(10.35) are the desired results for the problem of type B.

10.7 Special Cases and Applications

The complete system of operator equations introduced in previous sections may seem to be unfamiliar and somewhat abstract. To present them in a more familiar form and also to give us a partial check of our results, we shall reduce some of the above operator equations to a set of functional equations. Also, a brief discussion on systems with various reflecting operators $\mathbf{K}$ is presented.

10.7.1 External illumination

To discuss in more detail the generalized Chandrasekhar radiative problem in spherical geometry, the problem of type A, we assume that the core has an arbitrary reflecting kernel $K(v, u)$. The conical flux of radiation of unit intensity per unit area per unit solid angle is assumed to be spherically uniformly incident on the outer surface with radius y at inclination $\cos^{-1} u$, $0 < u \leq 1$, to the inward-directed radius vector, i.e.,

$$I_i = \delta(v - u).$$

To obtain the reflection functional equation at y, we let operators on both sides of eq. (10.20c) operate on I_i under our integral convention; see eq. (10.4). With the aid of eqs. (10.13), (10.16) and (10.17), we have

$$
\left[\frac{\partial}{\partial y} + \alpha(y) \left(\frac{1}{u} + \frac{1}{v} \right) + \frac{1 - v^2}{yv} \frac{\partial}{\partial v} + \frac{1 - u^2}{yu} \frac{\partial}{\partial u} - \frac{1 + u^2}{yu^2} \right] \rho(x, y, v, -u)
$$

$$
= \frac{\sigma}{2} \left(p(y, v, -u) + \int_0^1 p(y, v, w) \rho(x, y, w, -u) dw \right.
$$

$$
+ v \int_0^1 \rho(x, y, v, -w) p(y, -w, -u) \frac{dw}{w} \Bigg) \tag{10.36}
$$

$$
+ v \int_0^1 \int_0^1 \rho(x, y, -v, -w) p(y, -w, w') \rho(x, y, w', u) dw' \frac{dw}{w},
$$

with $\rho = 0$ when $x = y$. Likewise, using eqs. (10.10) and (10.12), eq. (10.26) is reduced to

$$
\rho(K; y, v, -u) = \rho(y, v, -u) + e^{-2h(x,y,u)} \left(\frac{y}{x} \right)^2 \frac{u}{u^*} k(K, v^*, -u_*)
$$

$$
+ e^{-h(x,y,u)} \left[\int_0^1 k(K, v^*, -w) \tau(x, y, -w, -u) \frac{dw}{w} \right.
$$

$$
+ \left(\frac{y}{x} \right)^2 \frac{u}{u^*} \int_0^1 t(x, y, v, w) k(K, w, -u^*) dw \Bigg] \tag{10.37}
$$

$$
+ \int_0^1 \int_0^1 t(x, y, v, w) k(K, w, -w') \tau(x, y, -w', -u) dw' \frac{dw}{w},
$$

where $\rho(K, x, y, v, -u)$ is the kernel for $\rho(K)$, $u^* = u^*(x, y)$, $u_* = u_*(x, y)$ and the resolvent kernel satisfies

$$
k(K, v, -u) \quad = K(v, -u)
$$

$$
+ \int_0^1 \int_0^1 k(K, v, -w) r(x, y, -w, w') \tag{10.38}
$$

$$
\cdot K(x, y, w', -u) dw' \frac{dw}{w}
$$

with $K(v, -u)$ as the kernel for **K**.

The desired intensity at the surface of the spherical shell of atmosphere in the outward direction is given by

$$I^d(y, u) = \rho(K, x, y, v, -u). \tag{10.39}$$

The corresponding functional equations for eqs. (10.20d) and (10.30) can be obtained by a similar method. They are, respectively,

$$\left(\frac{\partial}{\partial y} + \frac{\alpha(y)}{u} + \frac{1 - u^2}{yu}\frac{\partial}{\partial u} - \frac{1 + u^2}{yu^2}\right)\tau(x, y, -v, -u)$$

$$= \frac{\sigma}{2}\left[\frac{1}{v_*}e^{-h(x,y,u)}\left(p(y, -v_*, -u)\right.\right.$$

$$+ \int_0^1 p(y, -v_*, w)\rho(x, y, w, -u)dw\right) \tag{10.40}$$

$$+ \int_0^1 \tau(x, y, -v, -w)p(y, -w, -u)\frac{dw}{w}$$

$$+ \int_0^1\int_0^1 \tau(x, y, -v, -w)p(y, -w, w')\rho(x, y, w', -u)dw'\frac{dw}{w}\right],$$

with initial value $\tau = 0$ when $x = y$, and

$$\tau(K, x, y, -v, -u) = e^{-h(x,y,u)}(y/x)^2(u/u_*)k(K, -v, -u^*)$$
$$+ \int_0^1 k(K, -v, w)\tau(x, y, -w, -u)dw, \tag{10.41}$$

where $\tau(K, x, y, -v, -u)$ is the kernel of the operator $\tau(K)$. The radiation fields as seen by an observer at the surface of the core looking at the atmosphere are given by the following specular and diffuse parts,

$$I^s(x, -v) = \delta(v_* - u)e^{-h(x,y,u)} \tag{10.42}$$

and

$$I^d(x, -v) = \tau(K, x, y, -v, -u). \tag{10.43}$$

For the "standard problem" in a spherically symmetric shell, the core is a perfect absorber $K = 0$. Then

$$\rho(K, x, y, v, -u) = \rho(x, y, v, -u)$$

and

$$\tau(K, x, y, -v, -u) = \tau(x, y, -v, -u).$$

For the case $K = V$, eq. (10.38) reduces to

$$k(K, v, -u) \quad = \delta(v - u) + r(x, y, -v, u)$$
$$+ \int_0^1 r(x, y, -v, w) r(x, y, -w, u) dw + \cdots. \tag{10.44}$$

One may use eqs. (10.37), (10.41) and (10.44) to obtain $\rho(V, x, y, v, -u)$ and $\tau(V, x, y, -v, -u)$.

For Chandrasekhar's problem in a spherically symmetric shell, i.e., with $\mathbf{K}$ according to Lambert's law with a constant albedo A,

$$[\mathbf{K} \cdot I(x, -u)](v) = Av \int_0^1 uI(x, -u) dn \tag{10.45}$$

for any $I(x, -u)$ for all v, $0 < v \le 1$. The reflected radiation field is isotropic. By using Neumann's series

$$\mathbf{K} \cdot (\mathbf{E} - r \cdot \mathbf{K})^{-1} = \mathbf{K} + \mathbf{K} \cdot r \cdot \mathbf{K} + \mathbf{K} \cdot r \cdot \mathbf{K} \cdot r \cdot \mathbf{K} + \cdots$$

and repeating the operation (10.45) by taking advantage of the isotropic property, we obtain

$$I^d(y, v) = \rho(K) \cdot I_i = \rho(x, y, v, -u) + (1/vu)\bar{A}\underline{t}(v)\bar{\tau}(u),$$

$$I^d(x, -v) = \tau(K)I_i = (1/u)\bar{A}\bar{\tau}(u) + (\bar{A} - 1)(y/x)^2 \tag{10.46}$$

$$\times (u/u_*)e^{-h(x,y,u)},$$

where

$$\bar{A} = A(1 - A\bar{r})^{-1},$$

and

$$\underline{f}(v) \quad = \int_0^1 v f(x, y, v, u) du,$$
$$\bar{f}(u) \quad = \int_0^1 v f(x, y, -v, -u) dv. \tag{10.47}$$

$II^*(x, -v)$ has the same value as in eq. (10.42), since it is independent of $\mathbf{K}$. In summary, the results presented in eqs. (10.39), (10.42) and (10.43) are solutions for the general spherical shell of atmosphere with an arbitrary reflecting core and with external illumination. In the case $\mathbf{K}$ given by Lambert's law, we have the Chandrasekhar's radiation problem in spherical geometry. The results are presented in eqs. (10.46) and (10.47).

10.7.2 Internal illumination

As for the problem of type B, with internal illumination, $I_i = \delta(v - u)$ at x in the outward direction, we shall write down the functional equations from operator-equations (10.20a) and (10.20b) with the aid of eqs. (10.13), (10.16), (10.17) and (10.18). The result is

$$\left(\frac{\partial}{\partial y} + \frac{\sigma(y)}{v} + \frac{1-v^2}{yv}\frac{\partial}{\partial v}\right)t(x, y, v, u)$$

$$= \frac{\sigma(y)}{2}\left[\left(\frac{x}{y}\right)^2 e^{-h(x,y,u)}\left(p(y, v, u_*)\right.\right.$$

$$+ \int_0^1 \rho(x, y, v, -w)p(y, -w, u)\frac{dw}{w}\bigg)$$

$$+ \int_0^1 p(y, v, w)t(x, y, w, u)dw$$

$$+ \int_0^1 \int_0^1 \rho(x, y, v, -w)p(y, -w, w')t(x, y, w', u)dw'\frac{dw}{w}\bigg]$$

(10.48)

and

$$\frac{\partial}{\partial y}\tau(x, y, -v, u) = \left(\frac{x}{y}\right)^2\left(\frac{u}{u_*}\right)\left(e^{-2h(x,y,u)}p(y, -v_*, u^*)\right.$$

$$+e^{-h(x,y,u)}\int_0^1 \tau(x, y, -v, -w)p(y, -w, u_*)\frac{dw}{w}\bigg)$$

$$+e^{-h(x,y,u)}\int_0^1 p(y, -v_*, w)t(x, y, w, u)dw$$

$$+ \int_0^1 \int_0^1 \tau(x, y, -v, -w)p(y, -w, w')t(x, y, w', u)dw'\frac{dw}{w}$$

(10.49)

with initial conditions $r = t = 0$ at $x = y$. By using eqs. (10.32) and (10.33) the specular and transmitted intensities at y in the outward direction are

$$I^s(y, v) = \delta(v^* - u)e^{-h(x,y,u)},$$

$$I^d(y, v) = -[\delta(v^* - u) - \tilde{k}(K, v^*, -u)]e^{-h(x,y,u)}$$

$$+ \int_0^1 t(x, y, v, w)\tilde{k}(K, w, -u)dw$$

(10.50)

where $\tilde{k}(K, -v, u)$ is the modified resolvent which satisfies the equation

$$\tilde{k}(K, v, -u) = \delta(v - w)$$
$$+ \int_0^1 \int_0^1 K(v, -w)r(x, y, -w, w')k(V, w, -u)dw' \frac{dw}{w},$$

where $k(V, w, -u)$ satisfies (10.38) with $\mathbf{K} = \mathbf{V}$ or $K(v, -u) = \delta(v + u)$. The reflected intensity seen by the observer on the core surface looking into the atmosphere is, by eqs. (10.34) and (10.35),

$$I^d(x, -v) = \int_0^1 r(x, y, -v, w)\tilde{k}(K, w, -u)dw. \qquad (10.51)$$

For the "Schuster's problem," the core is an emitter and a perfect absorber. In this case, $r(K, x, y, -v, u) = r(x, y, -v, u)$ and $t(K, x, y, v, u) = t(x, y, v, u)$. For $I_i = \delta(v - u)$, answers for this problem are given by

$$I^s(y, v) \quad = \quad \delta(v^* - u)e^{-h(x, y, u)},$$

$$I^d(y, v) \quad = \quad t(x, y, v, u),$$

and

$$I^d(x, -v) = r(x, y, -v, u),$$

where t and r are given by eqs. (10.48) and (10.49).

For Milne's problem, $\tilde{k}(K, v, -u) = k(K, v, -u)$ has the expression as given in eq. (10.44). If $I_i = \delta(v - 1)$, i.e., the core radiates intensities in the normal direction. This is equivalent to the physical problem of the transfer of light from a central star surrounded by a spherical planetary nebula. Since incident radiation is normal to the inner surface of the spherical shell, $u = u^* = u_*$ and $v = v^* = v_*$.

For $\mathbf{K}$ given by Lambert's law, as discussed in the problems of type A, the output under $\mathbf{K}$ is isotropic and K has kernel Au. All integrations can be easily computed, for $I_i = \delta(v - u)$. The results are

$$I^s(y, v) \quad = \delta(v^* - u)e^{h(x, y, u)},$$

$$I^d(y, v) \quad = \mathbf{r}(K) \cdot I_i = (1/u)\bar{A}\bar{r}(u),$$

and

$$I^d(y, -v) = \mathbf{t}(K) \cdot I_i = (1/v)\bar{A}\underline{t}(v) + (\bar{A} - 1)\delta(v^* - u)e^{-h(x, y, u)},$$

where functions $\bar{A}$, $\bar{r}$, $\underline{t}$ are related to A, r, t; see (10.47).

For problem type A with external illumination I_i, the intensities at both sides of the spherical shell with radii x and y are related by

$$I(y, v) = (\mathbf{P} + \mathbf{t}) \cdot I(x, u) + \rho \cdot I_i, \qquad (10.52)$$

$$I(x, -v) = \mathbf{r} \cdot I(x, u) + (\mathbf{Q} + \tau) \cdot I_i. \tag{10.53}$$

From eq. (10.52),

$$(\mathbf{P} + \mathbf{t})I(x, u) = I(y, v) - \rho \cdot I_i = [\rho(K) - \rho] \cdot I_i.$$

By the results of eq. (10.26) and the fact that $(\mathbf{P} + \mathbf{t})$ is not a zero operator, we have

$$I(x, u) = \mathbf{K} \cdot (\mathbf{E} - \mathbf{r} \cdot \mathbf{K})^{-1}(\mathbf{Q} + \tau) \cdot I_i. \tag{10.54}$$

Upon substitution of eq. (10.54) into (10.53), using Neumann's series, we obtain

$$I(x, -v) = (\mathbf{E} - r \cdot K)^{-1} \cdot (\mathbf{Q} + \tau) \cdot I_i,$$

which agrees with eq. (10.28).

10.8 Numerical Solution of Functional Equations for Spherical Geometry

In earlier chapters of this book, we used quadrature techniques to approximate the integrals which appear in the Cauchy systems for reflection, transmission, internal intensity, source, and other functions of radiative transfer. Then we solved the approximate systems of ordinary differential equations to a high degree of precision. The Cauchy systems for spherical geometry presented in this chapter are more complicated, having several partial derivatives. Due to well-known difficulties in solving partial differential equations, we desire to reduce our problem to one involving only ordinary differential equations. We can do this in several ways. We describe a method for numerically estimating the partial derivatives, and alternatively, we present a technique for perturbation analysis. Then we implement these methods and present the results obtained for reflected intensity patterns.

The case we consider here is that of conical flux of net intensity π per unit area incident uniformly on a shell of inner radius a and outer radius z. The intensity of diffusely reflected radiation in the direction $\arccos v$, due to the incident flux with direction $\arccos u$, is

$$\rho(x, v, u) = S(z, v, u)/4v, \tag{10.55}$$

where the S function satisfies the functional equation,

$$\frac{\partial S(z, v, u)}{\partial z} + \frac{1 - v^2}{vz}\frac{\partial S}{\partial v} + \frac{1 - u^2}{uz}\frac{\partial S}{\partial u} + \left(\frac{1}{v} + \frac{1}{u}\right)S - \frac{v^2 + u^2}{v^2 u^2}\frac{S}{z}$$
$$= \lambda\left[1 + \frac{1}{2}\int_0^1 S(z, v, u')\frac{du'}{u'}\right]\left[1 + \frac{1}{2}\int_0^1 S(z, v', u)\frac{du'}{u'}\right], \quad z > a, \tag{10.56}$$

for $0 < v, u \le 1$, $z \ge a$. Supposing that the core is a perfect absorber, we set the initial condition,

$$S(a, v, u) = 0. \tag{10.57}$$

10.9 Numerical Estimation of Derivatives

We retain the Gaussian quadrature method for the evaluation of the integrals in the functional equation for S. Now we eliminate the partial derivatives by using linear combinations of the values of the functions at other points in the interval. The result is a system of equations for computing elements of the S matrix, which is similar in form to the system obtained for the slab case.

Let $x_1 < x_2 < \cdots < x_N$ be N points in an x-interval, and suppose that we wish to approximate the derivatives of a function $f(x)$ at the points x_i by linear combinations of the values of $f(x)$ at the x_i, $i = 1, 2, \ldots, N$,

$$f'(x_i) \cong \sum_{j=1}^{N} \alpha_j^{(i)} f(x_j). \tag{10.58}$$

Let us determine the coefficients, by analogy with the quadrature case, by the condition that the equations be exact for all polynomials of degree $N - 1$ or less. Using the trial functions $f(x) = x^k$, $k = 0, 1, \ldots, N - 1$, we obtain a system of linear algebraic equations

$$\sum_{j=1}^{N} x_j^{k-1} \alpha_j^{(i)} = (k - 1) x_i^{k-2}, \quad k = 1, 2, \ldots, N. \tag{10.59}$$

If we choose the x_i to be the N roots of the shifted Legendre polynomial of degree N, $\varphi_N(x)$, we can readily invert the coefficient matrix, a Vandermonde matrix [6].

Alternatively, we can obtain the $\alpha_j^{(i)}$ explicitly by using the N test functions [7],

$$f_i(x) = \frac{\varphi_N(x)}{x - x_i}. \tag{10.60}$$

The matrices $(\alpha_j^{(i)})$ for $N = 7$ and 9 are given in Tables 10.1 and 10.2.

Using quadrature on the integral terms and the foregoing approximations for the partial derivatives, eq. (10.56) is replaced by

$$\frac{dS_{ij}(z)}{dz} + \frac{1 - v_i^2}{v_i z} \sum_{k=1}^{N} \alpha_k^{(i)} S_{kj} + \frac{1 - v_j^2}{v_j z} \sum_{k=1}^{N} \alpha_k^{(j)} S_{ik} + \left(\frac{1}{v_i} + \frac{1}{v_j} \right) S_{ij}$$

$$- \frac{v_i^2 + v_j^2}{v_i^2 v_j^2} \frac{S_{ij}}{z} = \lambda \left[1 + \frac{1}{2} \sum_{k=1}^{N} S_{ik} \frac{w_k}{v_k} \right] \left[1 + \frac{1}{2} \sum_{k=1}^{N} S_{kj} \frac{w_k}{v_k} \right], \tag{10.61}$$

$z \geq a$, with initial conditions $S_{ij}(a) = 0$, $i, j = 1, 2, \ldots, N$.

Table 10.1. The Coefficients $\alpha_j^{(i)}$ for $N = 7$.

$i = 1$			
-0.19136364E 02	0.30166068E 02	-0.18345136E 02	0.12020668E 02
-0.73554054E 01	0.37037909E 01	-0.10536210E 01	
$i = 2$			
-0.30774001E 01	-0.32947313E 01	0.94826608E 01	-0.49141384E 01
0.27743267E 01	0.13485609E 01	0.37784329E-00	
$i = 3$			
0.73878691E 00	-0.37433740E 01	-0.97174703E 00	0.56413488E 01
-0.24639939E 01	0.10951929E 01	-0.29621352E-00	
$i = 4$			
-0.36940283E-00	0.14803137E 01	-0.43048331E 01	-0.99475983E-13
0.43048331E 01	-0.14803137E 01	0.36940283E-00	
$i = 5$			
0.29621352E-00	-0.10951929E 01	0.24639939E 01	-0.56413488E 01
0.97174703E 00	0.37433740E 01	-0.73878691E 00	
$i = 6$			
-0.37784329E-00	0.13485609E 01	-0.27743267E 01	0.49141384E 01
-0.94826608E 01	0.32947313E 01	0.30774001E 01	
$i = 7$			
0.10536210E 01	-0.37037909E 01	0.73554054E 01	-0.12020668E 02
0.18345136E 02	-0.30166068E 02	0.19136364E 02	

10.10 Perturbation Approximation

Let us now turn our attention to a power series expansion technique that allows the reduction of the original problem to a set of problems similar to that of the parallel slab case.

Write $x = z - a$, assume that $x/a \ll 1$, and set

$$S = S_0 + \frac{S_1}{a} + \frac{S_2}{a^2} + \cdots, \tag{10.62}$$

where $S_0, S_1, S_2, \ldots$ are independent of a. Substituting in (10.56) and equating coefficients, we obtain the following equations:

$$(S_0)_x + \left(\frac{1}{u} + \frac{1}{v}\right) S_0$$
$$= \lambda \left[1 + \frac{1}{2} \int_0^1 S_0(x, v', u) \frac{dv'}{v'}\right]\left[1 + \frac{1}{2} \int_0^1 S_0(x, v, u') \frac{du'}{u'}\right], \tag{10.63}$$

$$(S_1)_x + \left(\frac{1 - v^2}{v}\right)(S_0)_v + \left(\frac{1 - u^2}{u}\right)(S_0)_u + \left(\frac{1}{u} + \frac{1}{v}\right) S_1 - \frac{u^2 + v^2}{u^2 v^2} S_0$$
$$= \lambda\left[1 + \frac{1}{2}\int_0^1 S_0(x, v, u')\frac{du'}{u'}\right]\left[\frac{1}{2}\int_0^1 S_1(x, v', u)\frac{dv'}{v'}\right]$$
$$+ \lambda\left[1 + \frac{1}{2}\int_0^1 S_0(x, v', u)\frac{dv'}{v'}\right]\left[\frac{1}{2}\int_0^1 S_1(x, v, u')\frac{du'}{u'}\right]. \tag{10.64}$$

Table 10.2. The Coefficients $\alpha_j^{(i)}$ for $N = 9$.

$i = 1$			
-0.30899183E 02	0.49462602E 02	-0.31847722E 02	0.23009713E 02
-0.16634325E 02	0.11463908E 02	-0.71444862E 01	0.36223711E 01
-0.10328869E 01			
$i = 2$			
-0.46321847E 01	-0.55540647E 01	0.15529632E 02	-0.88594615E 01
0.58950087E 01	-0.39077266E 01	0.23856884E 01	-0.11961277E 01
0.33923594E-00			
0.33923594E-00			
$i = 3$			
0.99779608E 00	-0.51953604E 01	-0.19666417E 01	0.90706996E 01
-0.46474057E 01	0.27969636E 01	-0.16303335E 01	0.79812006E 00
-0.22383800E-00			
$i = 4$			
-0.41927865E-00	0.17238123E 01	-0.52755643E 01	-0.72470224E 00
0.67044574E 01	-0.30840075E 01	0.16267280E 01	-0.76033820E 00
0.20889316E-00			
$i = 5$			
0.25654308E-00	-0.97080200E 00	0.22877170E 01	-0.56744949E 01
0.56843419E-11	0.56744949E 01	-0.22877170E 01	0.97080200E 00
-0.25654308E-00			
$i = 6$			
-0.20889316E-00	0.76033820E 00	-0.16267280E 01	0.30840075E 01
-0.67044574E 01	0.72470224E 00	0.52755643E 01	-0.71238123E 01
0.41927865E-00			
$i = 7$			
0.22383800E-00	-0.79812006E 00	0.16303335E 01	-0.27969636E 01
0.46474057E 01	-0.90706996E 01	0.19666417E 01	0.51953605E 01
-0.99779608E 00			
$i = 8$			
-0.33923594E-00	0.11961277E 01	-0.23856884E 01	0.39077266E 01
-0.58950087E 01	0.88594615E 01	-0.15529632E 02	0.55540647E 01
0.46321847E 01			
$i = 9$			
0.10328869E 01	-0.36223711E 01	0.71444762E 01	-0.11463908E 02
0.16634325E 02	-0.23009713E 02	0.31847722E 02	-0.49462602E 02
0.30899183E 02			

Using the quadrature techniques described above, we reduce these equations to a system of ordinary differential equations with initial conditions and readily obtain numerical solutions. In Section 10.11 we shall compare computational results obtained in different fashions.

If the thickness of the shell is small compared with the inner radius, we can expect this perturbation technique to provide excellent results. If $x \geq a$, we face divergence.

There are several ways of overcoming this difficulty. One method is to do the calculation in parts. First, we carry it out for $x/a \leq 0.1$, say. Then we consider a new problem in which the inner radius is a. This replaces the complete absorber by an inhomogeneous reflecting material, but this is not a matter of any difficulty. It merely yields a new initial condition. We can proceed in this fashion step by step until we obtain the desired shell thickness.

Another approach is based upon the observation that the divergence of the power series for $|x| \geq a > 0$:

$$\frac{1}{a+x} = \frac{1}{a}\left[1 - \frac{x}{a} + \frac{x^2}{a^2} - \cdots\right] \tag{10.65}$$

is due to the singularity at $x = -a$. However, we are interested only in $x \geq 0$. Let us then set

$$y = \frac{x}{x+k}, \qquad x = \frac{ky}{1-y} \tag{10.66}$$

for some suitably chosen k and expand in powers of y. Thus,

$$\frac{1}{a+x} = \frac{1}{a+ky/1-y} = \frac{1-y}{a+(k-a)y}. \tag{10.67}$$

A convenient choice is $k = a$. Although we do not know the analyticity properties of S as a function of x, we do know that the function exists for $x \geq 0$ and that $0 \leq y \leq 1$ for $x \geq 0$. A detailed discussion of this device for analytic continuation with further references and applications will be found in [6].

10.11 Numerical Results

The above procedures for the calculation of S are carried out with FORTRAN source programs. In the first series of numerical experiments, we produce the S function for a shell by integrating the system of differential equations (10.61). We call this Method I. We use formulas of order $N = 7$ and $N = 9$, and integration step sizes of 0.005 and 0.0025. There is agreement among calculations for comparable cases.

With $N = 7$ and a step size of 0.005, we vary the inner radius of the shell, $a = 100, 500,$ and 1000. We compare reflected intensities, $r = S/4v$, for the shell against the corresponding intensities for the plane-parallel slab, which should be obtained as $a \to \infty$. The results are shown in Fig. 10.6. The reflection function r is shown for the case in which the albedo is 1 and the thickness is 3 for three angles of incidence that are approximately 13.0°, 60.0°, and 88.5°. We see immediately that the curves for the shell geometry always lie on or above the curves for the slab. In particular, the curve for 88.5°, with $a = 100$, lies as much as 50% above the curve for the slab. As the inner radius a is increased, the r function for the shell approaches that for the slab. The two cases are graphically indistinguishable for $a = 1000$. For the angle of incidence 60°, we have drawn a dashed curve for $a = 50$. It

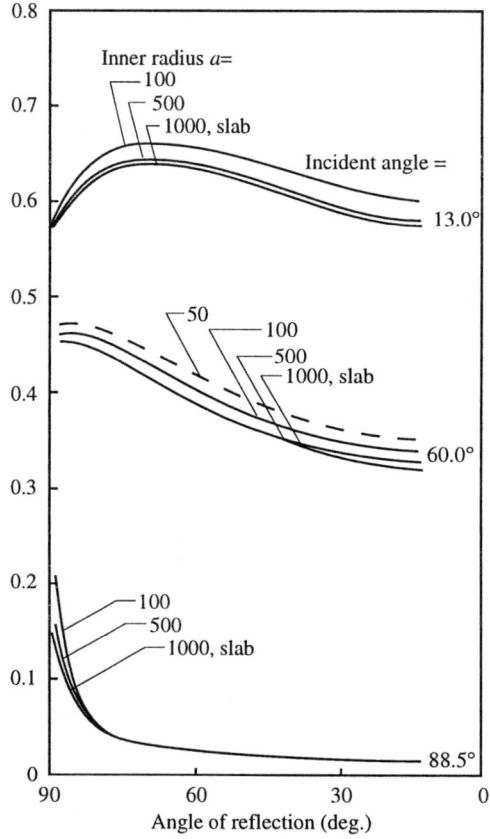

Figure 10.6. Some reflected intensity patterns for shells with albedo $\lambda = 1$ and thickness $x = 3$, for various angles of incidence.

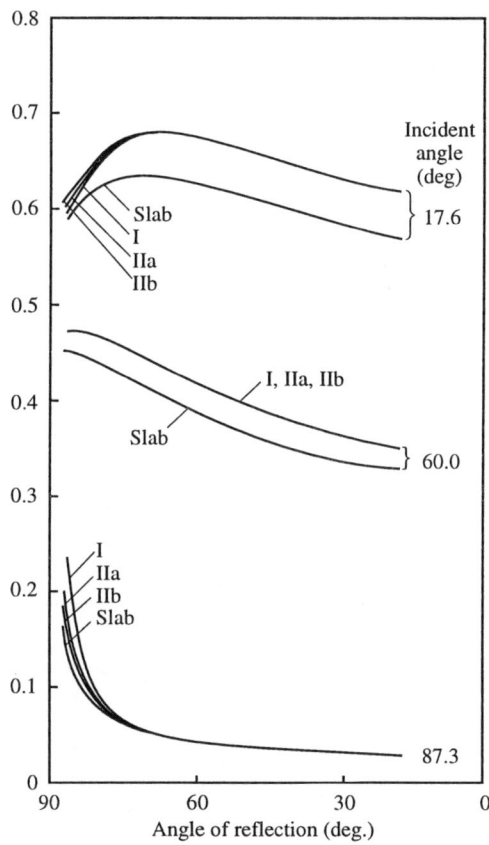

Figure 10.7. Some reflected intensity patterns for a shell with albedo $\lambda = 1$, inner radius $a = 50$, and thickness $x = 3$, for various angles of incidence.

is the result of a calculation with $N = 5$, since the calculation for $N = 7$ is unstable.

Computations of the reflection function r are also carried out with variations of the perturbation technique. The partial derivatives $(S_0)_v$ and $(S_0)_u$ which appear in (10.64) are in the first instance approximated by formula (10.58) in Method IIa, and secondly are produced as solutions of systems of differential equations in Method IIb. Checks consisting of varying the order of the quadrature formula, varying the step length of integration, and increasing the inner radius are positive.

A comparison of the results of Methods I, IIa, and IIb shows satisfactory agreement. Figure 10.7 shows three sets of curves for the reflection function r, for the case in which the albedo is 1, the inner radius is 50, and the thickness is 3. Each set corresponds to a different angle of incidence, 17.6°, 60.0°, and 87.3°. The order of the quadrature formula is $N = 5$. Four curves are plotted for each angle, although in some instances they lie on top of one another. These curves are labeled "Slab," "I," "IIa," and "IIb," in an obvious notation.

Computing times for the different methods are short. The perturbation technique, however, has the advantage of producing reflection functions for a variety of shell inner radii in a single calculation. This technique is found to be stable, and gives good results even when the ratio x/a is fairly large. This ratio is 3/50 for the case represented by Fig. 10.7.

10.12 Discussion

In this chapter, we used the invariant imbedding method to construct a set of reflection and transmission operators for a spherical shell. These operators yield the invariant imbedding functional equations. Except for the assumption that the incident radiation is spherically uniform, our result is complete and general. It includes inhomogeneity and anisotropy, allows for internal or external illumination, and treats both reflecting and absorbing cores.

The computational aspects of solving the functional equations were considered. The aim, as usual, was to transform the partial differential equations to initial value problems. We kept the derivative in the radial direction as the independent variable. We used Gaussian quadrature to approximate the definite integrals over direction cosines. The partial derivatives with respect to the direction cosines were handled in several ways. In a perturbation approach, those partial derivatives dropped out of the picture and a system of ordinary differential equations was obtained for the terms in the expansion. In another approach, the partial derivatives were approximated as linear combinations of the values of the function at the same points as used in the quadrature; then a system of ordinary differential equations resulted. Alternatively, these partial derivatives themselves were regarded as solutions of

functional equations of similar form, and they were determined simultaneously through integrating an appropriate system of differential equations.
These computations were carried out and the results analyzed, with the conclusion that the methods are quite useful and accurate over a wide range.
The foregoing experiments were carried out for the homogeneous shell with
isotropic scattering.

The analytical results of this chapter plus the computational work leads
us to believe that more general cases of spherical symmetry can be studied
computationally. Of course, computational experiments must be carried out
systematically in order to make steady advances.

References

1. R. E. Bellman, H. H. Kagiwada and R. E. Kalaba, "Invariant Imbedding and Perturbation Techniques Applied to Diffuse Reflection from Spherical Shells," *The Rand Corporation*, RM-4730-NASA, August 1965.
2. R. Bellman, H. Kagiwada and R. Kalaba, "Invariant Imbedding and Radiative Transfer in Spherical Shells," *J. Comput. Phys.*, Vol. 1, 1966, pp. 245–256.
3. R. Bellman, H. Kagiwada, R. Kalaba and S. Ueno, "Diffuse Reflection of Solar Rays by a Spherical Shell Atmosphere," *Icarus*, Vol. 11, 1969, pp. 417–423.
4. R. Bellman, H. Kagiwada, R. Kalaba and S. Ueno, "Diffuse Transmission of Light from a Central Source Through an Inhomogeneous Spherical Shell with Isotropic Scattering," *The Rand Corporation*, RM-5216-PR; *J. Math. Phys.*, Vol. 9, 1966, pp. 909–912.
5. Alan P. Wang, "Linear Operators and Transfer of Radiation with Spherical Symmetry," *J. Math. Physics*, Vol. 14, 1973, pp. 855–862.
6. R. Bellman, I. Glicksberg and O. Gross, "Some Aspects of the Mathematical Theory of Control Processes," *The RAND Corporation*, R-313, 1958.
7. R. Bellman, H. Kagiwada, R. Kalaba and M. Prestrud, *Invariant Imbedding and Time-Dependent Processes*, Elsevier, New York, 1964.
8. R. Bellman, H. Kagiwada, R. Kalaba and S. Ueno, "A New Derivation of the Integro-Differential Equation for Chandrasekhar's X and Y Function," *J. Math. Physics*, Vol. 9, 1968, pp. 906–908.
9. H. Kagiwada, R. Kalaba and R. Bellman, "Numerical Estimation of Derivatives with an Application to Radiative Transfer in Spherical Shells," *The Rand Corporation*, RM-4617-NASA, June 1965.
10. R. Bellman and R. Kalaba, *Proc. Natl. Acad. Sci. U.S.*, Vol. 54, 1965, pp. 1293–1296.
11. S. Ueno, R. Bellman, H. Kagiwada and R. Kalaba, "The Spectral Brightness of an Inhomogeneous Spherical Planetary Atmosphere," in *Space Research IX*, KS. W. Champion, P. A. Smith and R. L. Smith-Rose, eds., North-Holland, Amsterdam, 1969, pp. 385–391.
12. S. Ueno, H. Kagiwada and R. Kalaba, "Radiative Transfer in Spherical Shell Atmospheres with Radial Symmetry," *J. Math. Phys.*, Vol. 12, 1971, pp. 1279–1286.

11. Bibliography

I. BOOKS

1.   R. Bellman, H. Kagiwada, R. Kalaba and M. Prestrud, *Invariant Imbedding and Time-dependent Transport Processes*, American Elsevier Publishing Co., New York, 1963.

2.   R. Bellman and G. M. Wing, *An Introduction of Invariant Imbedding*, SIAM, Philadelphia, 1962.

3.   C. F. Bohren, ed., *Selected Papers on Scattering in the Atmosphere*, SPIE Optical Engineering Press, Bellingham, Washington, 1989.

4.   I. W. Busbridge, *The Mathematics of Radiative Transfer*, Cambridge University Press, Cambridge, 1960.

5.   S. Chandrasekhar, *Radiative Transfer*, Dover Publications, New York, 1960.

6.   R. M. Goody and Y. L. Yung, *Atmospheric Radiation: Theoretical Basis*, Oxford University Press, New York, 1989.

7.   R. P. Gupta, *Remote Sensing Geology*, Springer Verlag, New York, 1991.

8.   Y. Haimes and J. Kindler, eds., *Water and Related Land Resource Systems*, Pergamon Press, New York, 1980.

9.   H. Kagiwada, *System Identification: Methods and Applications*, Addison-Wesley Publishing Co., Reading, Mass., 1974.

10.   H. Kagiwada, R. Kalaba, N. Rasakhoo and K. Spingarn, *Numerical Derivatives and Nonlinear Analysis*, Plenum Press, New York, 1986.

11.   H. Kagiwada, R. Kalaba and S. Ueno, *Multiple Scattering Processes: Inverse and Direct*, Addison-Wesley Publishing Co., Reading, Mass., 1975.

12.   W. Kalkofen, ed., *Numerical Radiative Transfer*, Cambridge Univ. Press, Cambridge, 1988.

13.   K. Y. Kondratyev, *The Atmospheric Effect in the Investigation of Natural Resources Coming from the Cosmos*, (in Russian), Moscow, Machinostroenie, 1985.

14.   K. Y. Kondratyev et al., *Remote Sensing of the Earth from Space: Atmospheric Correction*, Springer Verlag, New York, 1992.

15.   V. Kourganoff, *Basic Methods in Transfer Problems*, Dover Publications, New York, 1963.

16.   J. H. Kramer, *Earth Observation–Remote Sensing*, Springer Verlag, New York, 1992.

17.   J. Lenoble, ed., *Radiative Transfer in Scattering and Absorbing Atmospheres: Standard Computational Procedures*, A. Deepak Publishing, 1995.

18.   J. Lenoble and J. F. Geleyn, eds., *IRS 88: Current Problems in Atmospheric Radiation*, A Deepak Publishing, 1989.

19.   Kuo-Nan Liou, *An Introduction to Atmospheric Radiation*, Academic Press, New York, 1980.

20.   D. H. Menzel, ed., *Transfer of Radiation*, Dover Publications, New York, 1966.

21.   M. A. Mnatsakanian and H. V. Pickichian, eds., *Principle of Invariance and its Applications*, Publishing House of Academy of Sciences, Yerevan, Russian, 1989.

22.   J. A. Richards, *Remote Sensing Digital Image Analysis*, Springer Verlag, New York, 1993.

23.   V. V. Sobolev, *A Treatise on Radiative Transfer*, D. Van Nostrand Co., New York, 1963.

24.   V. V. Sobolev, *Light Scattering in Planetary Atmospheres*, Pergamon Press, New York, 1974.

25.   H. G. Van De Hulst, *Multiple Light Scattering: Tables, Formulas, and Applications*, Vol. 1 and 2, Academic Press, New York, 1980.

26.   F. J. Vernberg and F. P. Diemer, eds., *Processes in Marine Remote Sensing*, University of South Carolina Press, 1982.

27.   E. G. Yanovitskij, *Light Scattering in Inhomogeneous Atmospheres*, Springer-Verlag, New York, 1996.

II. PAPERS

1. L. A. Apresyan, "Method of Invariant Embedding for the Radiation Transfer Equation and Description of the Scattering of Finite Beams," *Radiophysics and Quantum Electronics*, Vol. 33, No. 9, 1990, pp. 766–772.

2. V. V. Barun, "Detection of a Small Target Against Bottom of Water Reservoir," *Proceedings of the SPIE–The International Society for Optical Engineering, SPIE*, Vol. 2759, 1996, pp. 490–501.

3. A. Ben-David, "Multiple-Scattering Transmission and an Effective Average Photon Path Length of a Plane-Parallel Beam in a Homogeneous Medium," *Applied Optics*, Vol. 34, No. 15, 1995, pp. 2802–2810.

4. R. Bellman, R. Kalaba and G. M. Wing, "Invariant Imbedding and Mathematical Physics I: Particle Processes," *Rand Corp.*, Report P-1858, 1960.

5. R. Bellman, R. Kalaba and S. Ueno, "Invariant Imbedding and Time-Dependent Diffuse Reflection by a Finite Inhomogeneous Atmosphere," *Icarus*, Vol. 1, 1962, pp. 191–199.

6. R. Bellman, H. Kagiwada, R. Kalaba and S. Ueno, "Inverse Problems in Radiative Transfer: Layered Media," *Icarus*, Vol. 4, 1965, pp. 119–126.

7. R. Bellman, H. Kagiwada, R. Kalaba and S. Ueno, "On the Identification of Systems and the Unscrambling of Data–II: An Inverse Problem in Radiative Transfer," *Proc. Nat. Acad. Sci.*, USA, Vol. 153, 1965, pp. 910–913.

8. R. Bellman, H. Kagiwada and R. Kalaba, "Invariant Imbedding and a Reformulation of the Internal Intensity Problem in Radiative Transfer Theory," *Monthly Notices of Royal Astronom. Society*, Vol. 132, 1966, pp. 183–191.

9. R. Bellman, H. Kagiwada and R. Kalaba, "Estimation of Internal Source Distributions Using External Field Measurements in Radiative Transfer," *Icarus*, Vol. 5, 1966, pp. 274–278.

10. R. Bellman, H. Kagiwada and R. Kalaba, "Invariant Imbedding and Radiative Transfer in Spherical Shells," *J. Comp. Physics*, Vol. 1, 1966, pp. 245–256.

11. A.M. Bruckstein and T. Kailath, "Inverse Scattering for Discrete Transmission-Line Models," *SIAM Review*, Vol. 29, No. 2, 1987, pp. 359–389.

12. J. Buell, H. Kagiwada, R. Kalaba, A. McNabb and A. Schumitzky, "Computation of the Resolvent for the Auxiliary Equations of Radiative Transfer," *J. Quant. Spect. Rad. Transfer*, Vol. 8, 1968, pp. 1481–1489.

13. E. Kh. Danielyan, "Theory of the Isotropic Scattering of Radiation in a Plane Layer: Feasibility of Obtaining a Complete Analytical Solution of the Problem," *Astrophysics*, Vol. 37, No. 1, 1994, pp. 79–88.

14. W. L. Dunn and C. E. Siewert, "The Searchlight Problem in Radiation Transport: Some Analytical and Computational Results," *Zeitschrift fur Angewandte Mathematik und Physik*, Vol. 36, No. 4, 1985, pp. 581–595.

15. B. D. Ganapol and D. W. Nigg, "Analytical Two-Dimensional Neutron Transport Benchmark: The Searchlight Problem," *Transactions of the American Nuclear Society*, Vol. 64, 1991, pp. 276–278.

16.   L. Gonima, "Simple Algorithm for the Atmospheric Correction of Reflectance Images," *International Journal of Remote Sensing*, Vol. 14, No. 6, 1993, pp. 1179–1187.

17.   H. R. Gordon and M. Wang, "Influence of Oceanic Whitecaps on Atmospheric Correction of Ocean-Color Sensors," *Applied Optics*, Vol. 33, No. 33, pp. 7754–7763.

18.   J. F. de Haan, J. W. Hovenier, J. M. M. Kokke, H. T. C. van Stokkom, "Removal of Atmospheric Influences on Satellite-Born Imagery: A Radiative Transfer Approach," *Remote Sensing of Environment*, Vol. 37, No. 1, 1991, pp. 1–21.

19.   A. J. Haines and M. V. de Hoop, "An Invariant Imbedding Analysis of General Wave Scattering Problems," *J. Math. Physics*, Vol. 37, No. 8, 1996, pp. 3854–3881.

20.   P. H. Hauschildt, "A Fast Operator Perturbation Method for the Solution of the Special Relativistic Equation of Radiative Transfer in Spherical Symmetry," *Journal of Quantitative Spectroscopy and Radiative Transfer*, Vol. 47, No. 6, 1992, pp. 433–453.

21.   H. Kagiwada, R. Kalaba, and S. Ueno, "Initial-Value Method for Integral Equation Arising in Theory of the Solar Atmosphere," *Astrofisika*, Vol. 4, Acad. Sci. Arm. SSR, 1968, pp. 498–503.

22.   H. Kagiwada and R. Kalaba, "Direct and Inverse Problems for Integral Equations Via Initial Value Methods," *SIAM-AMS Proc. on Transport Theory*, Vol. 1, 1969, pp. 112–128.

23.   H. Kagiwada and R. Kalaba, "Exact Solution of a Family of Integral Equations of Anisotropic Scattering," *J. Math. Phys.*, Vol. 11, 1970, pp. 1575–1578.

24.   H. Kagiwada and R. Kalaba, "Invariant Imbedding and Radiation Fields in Finite Isotropically Scattering Slabs Bounded by a Lambert's Law Reflector," *J. Quant. Spect. Rad. Transfer*, Vol. 11, 1971, pp. 1101–1109.

25.   H. Kagiwada, R. Kalaba and S. Ueno, "Radiative Transfer in Spherical Shell Atmospheres with Radial Symmetry," *J. Math. Physics*, Vol. 12, 1971, pp. 1279–1286.

26.   H. Kagiwada, R. Kalaba, S. Timko and S. Ueno, "Associative Memories for System Identification: Inverse Problems in Remote Sensing," in *Mathematical and Computer Modeling*, Vol. 14, 1990, pp. 200–202.

27.   H. H. Kagiwada, S. Ueno and Y. Kawata, "Simulation of the Atmospheric and Topographic Effects in Remote, Sensing from Space," in *Proc. of the IGARSS'90*, Washington, May 20-24, 1990, Vol. 1, pp. 183–186.

28.   H.H. Kagiwada, J. K. Kagiwada and S. Ueno, "Kalaba's Associative Memories for System Identification," in *Appl. Math. Comp.*, Elsevier Sci. Publ. Co., New York, Vol. 45, 1991, pp. 135–142.

29.   H. Kagiwada, S. Ueno and Y. Kawata, "Analytical Approximation of Atmospheric Correction in Rugged Terrain," *XVII ISPRS Congress*, Aug. 2-14, 1992, Washington, D.C.

30.   W. Kalkofen and R. Wehrse, "Radiative Transfer in an Expanding Spherical Medium," in *Proceedings of the Third Cambridge Workshop on Cool Stars, Stellar Systems and the Sun*, Springer-Verlag, Berlin, 1984.

31.   Y. Kawata, Y. Haba, T. Kusaka and S. Ueno, "The System of Correcting Remotely Sensed Earth's Imagery for Atmospheric Effects," in *Proc. of 13th Intl. Symp. on Remote Sensing of Environ.*, ERIM, Ann Arbor, MI, April 23-27, 1979, pp. 1883-1894.

32.   Y. Kawata, T. Kusaka and S. Ueno, "Landsat Imaging: Removal of Atmospheric Effects," in *Encyclopedia of Systems and Control*, M. Singh, ed., Pergamon Press, New York, 1987, pp. 2652-2655.

33.   Y. Kawata, H. Koide and S. Ueno, "On the Classification Accuracy for Images Obtained by Landsat, Spot and MOS-1 Sensors," *Preprints of the 20th ISCIE SSS Osaka*, Japan, Nov. 16-18, 1988, pp. 135-138.

34.   Y. Kawata, S. Ueno and A. Ohtani, "The Surface Albedo Retrieval of Mountainous Forest Area from Satellite MSS Data," in *Appl. Math. Comp.*, Vol. 69, 1995, pp. 41-60.

35.   V. V. Khutortsev, "Use of the Method of Invariant Embedding for Synthesizing an Iterative Algorithm for Finding an Optimal Control of a Measurement Process," *Journal of Computer and Systems Sciences International*, Vol. 30, No. 6, 1992, pp. 133-137.

36.   J. Lenoble, "Solution of Radiative Transfer in an Infinite Scattering Medium Illuminated by a Point Source," Scientific Report No. 6, Department of Meteorology, *University of California*, 1960.

37.   C. M. Leung, "Numerical Solution of the Radiative Transfer Equation in Spherically Symmetric Dust Shells," *Journal of Quantitative Spectroscopy and Radiative Transfer*, Vol. 16, No. 7, 1976, pp. 559-574.

38.   M. Matsumoto, "Multiple Scattering Processes in Non-Stationary Radiation Field," *Optical Review*, Vol. 2, No. 4, 1995, pp. 249-254.

39.   M. I. Mishchenko, "Multiple Scattering of Polarized Light in Anisotropic Plane-Parallel Media," *Transport Theory and Statistical Physics*, Vol. 19, Nos. 3-5, 1990, pp. 293-316.

40.   M. Missana, "Solution of the Transfer Equation in a Scattering Atmosphere with Spherical Symmetry," *Astrophysics and Space Science*, Vol. 33, No. 1, 1975, pp. 245-251.

41.   M. Moriyama and K. Arai, "Iterative Estimation of the Earth Surface Temperature and Emissivity," *Advances in Space Research*, Vol. 16, No. 10, 1995, pp. 117-120.

42.   P. Nelson, "Invariant Embedding and Related Methods," *Transactions of the American Nuclear Society*, Vol. 71, 1994, pp. 223-224.

43.   A. P. Odell and J. A. Weinman, "The Effect of Atmospheric Haze on Images of the Earth's Surface," *J. Geophys. Research*, Vol. 80, 1975, pp. 5035-5040.

44.   A. Peraiah and B. A. Varghese, "Radiative Transfer Equation in Spherical Symmetry," *Astrophysical Journal*, Vol. 290, No. 2, 1985, pp. 411-423.

45. A. M. Perez, P. Illera, and J. L. Casanova, "Analysis of Different Models for Atmospheric Correction of Meteosat Infrared Images. A New Approach," *Atmospheric Research*, Vol. 30, No. 1, 1993, pp. 1–12.

46. R. Redheffer, "On the Relation of Transmission Line Theory to Scattering and Transfer," *J. Mathematics and Physics*, Vol. XLI, No. 1, March 1962, pp. 1–41.

47. R. Redheffer and A. P. Wang, "Formal Properties of Time-Dependent Scattering Processes," *J. Math. Mech.*, Vol. 19, No. 9, 1970, pp. 765–781.

48. G. B. Rybicki, "The Searchlight Problem with Isotropic Scattering," *Journal of Quantitative Spectroscopy and Radiative Transfer*, Vol. 11, No. 6, 1971, pp. 827–849.

49. J. Schmid-Burgk, "Radiative Transfer Through Spherically-Symmetric Atmospheres and Shells," *Astronomy and Astrophysics*, Vol. 40, No. 3, 1975, pp. 249–255.

50. C. E. Siewert and W. L. Dunn, "On Inverse Problems for Plane-Parallel Media with Nonuniform Surface Illumination," *Journal of Mathematical Physics*, Vol. 23, No. 7, 1982, pp. 1376–1378.

51. S. M. Singh, "Fast Atmospheric Correction Algorithm," *International Journal of Remote Sensing*, Vol. 13, No. 5, 1992, pp. 933–938.

52. A. A. Strotsev and V. V. Khutortsev, "Optimal Active Experimental Design by Inverse Invariant Embedding," *Automation and Remote Control*, Vol. 54, No. 5, 1993, pp. 807–811.

53. R. E. Turner and M. N. Spencer, "Atmospheric Model for Correction of Spacecraft Data," in *Proceedings of Eighth International Symposium of Remote Sensing of Environment*, University of Michigan, Ann Arbor, 1972, Michigan, pp. 895–934.

54. R. E. Turner, "Signature Variations Due to Atmospheric Effects," in *Proceedings of the Tenth International Symposium on Remote Sensing of Environment*, University of Michigan, Ann Arbor, Michigan, 1975, pp. 671–682.

55. S. Ueno, "On the Diffusion Matrix of Radiative Transfer," *Ann. d' Astrophys.*, Vol. 24, 1961, pp. 352–358.

56. S. Ueno, "The Invariant Imbedding Method for Transport Problems. II. Resolvent in Photon Diffusion Equation," *J. Math. Anal. Appl.*, Vol. 3, 1961, pp. 361–372.

57. S. Ueno, "On the Time-Dependent Principle of Invariance in a Semi-Infinite Medium," *J. Math. Anal. Appl.*, Vol. 4, 1962, pp. 1-8.

58. S. Ueno, Y. Kawata, T. Kusaka and Y. Haba, "Ground Albedo Mapping from Remotely Sensed Earth's Imagery Data," in *Water and Related Land Resource System*, Y. Haimes and J. Kindler, eds., Pergamon Press, New York, 1980, pp. 211–218.

59. S. Ueno and Y. Kawata, "Optical Assessment of Subsurface Water Parameters Using Radiance Measurements from Space," Papers selected for

16th Intl. Symp. on Remote Sensing of Environ., Buenos Aires, Argentina, June, 1983, *ERIM*, pp. 409–421.

60.   S. Ueno, "Cauchy System for Resolvent of Milne's Integral Equations with Anisotropic Scattering," *J. Math. Phys.*, Vol. 26, 1985, pp. 85–88.

61.   S. Ueno, "Cauchy System for the Scattering Function in the Searchlight Problem," *J. Math. Anal. Appl.*, Vol. 119, 1986, pp. 223–233.

62.   S. Ueno and Y. Kawata, "Identification of Optical Parameters in Earth-Atmosphere System," *Encyclopedia of Systems and Control*, M. Singh, ed., Pergamon Press, New York, 1987, pp. 1307–1311.

63.   S. Ueno, Y. Kawata and T. Kusaka, "Ground Albedo Mapping: Invariant Embedding," *Encyclopedia of Systems and Control*, M. Singh, ed., Pergamon Press, New York, 1987, pp. 2046–2053.

64.   S. Ueno and A. P. Wang, "Invariant Imbedding and Searchlight Problem in Turbid Slab," *Computers Math. Appl.*, Vol. 21, 1991, pp. 1–6.

65.   S. Ueno, M. Toho and A. P. Wang, "Theoretical and Numerical Solutions of Searchlight Problem," in *Proc. of IGARSS'93*, Tokyo, Japan, Aug. 18-21, 1993, pp. 1279–1282.

66.   H. C. van de Hulst, "Radiative Transfer in a Spherical Dust Cloud, II. Asymptotic Form of the Reflection Function for Isotropic Scattering," *Astronomy and Astrophysics*, Vol. 207, No. 1, 1988, pp. 182–193.

67.   A. P. Wang, "Invariant Imbedding and Scattering Processes," *J. Math. Anal. Appl.*, Vol. 17, No. 1, 1967, pp. 48–60.

68.   A. P. Wang, "Discrete Radiative Transfer," *J. Math. Anal. Appl.*, Vol. 24, No. 3, 1968, pp. 530–544.

69.   A. P. Wang, "Nonstationary Multiple Scattering," *J. Math. Phys.*, Vol. 18, No. 1, 1977, pp. 47–51.

70.   A. P. Wang, "A Mathematical Model of a Cloud," *J. Astrophys. Space Science*, Vol. 70, 1980, pp. 447–459.

71.   A. P. Wang, "Correction of Atmospheric Effects in Remote Sensing," *Math. Modelling, An International Journal*, Vol. 9, No. 2, 1987, pp. 117–124.

72.   A. P. Wang and Steve Shaw, "Scattering Operators and Propagation Solutions for a Non-Stationary Transfer Equation, *J. Math. Anal. Appl.*, Vol. 134, No. 1, 1988, pp. 170–185.

73.   A. P. Wang and S. Ueno, "An Inverse Problem in a Three-Dimensional Radiative Transfer," *J. Astrophysics and Space Science*, Vol. 155, No. 1, 1989, pp. 105–111.

74.   A. P. Wang and S. Ueno, "Searchlight Problem in Turbid Slate," *Computer Math. Appl.*, Vol. 21, No. 11/12, 1991, pp. 1–6.

75.   A. P. Wang and S. Ueno, "Identification of a Transport System," to appear in *Proc. of Intl. Symposium on Artificial Life and Robotics*, Feb. 18-20, 1996, B-ConPlaza, Beppu, Oita, Japan, pp. 42–46.

A. Appendix A
The Physical Problem of Radiative Transfer

Radiative transfer is the study of the transfer of radiant energy in atmospheres. The examination of the physical processes of absorption and multiple scatterings leads to mathematical descriptions. These equations, when solved, yield the intensities of radiant energy, the rate of production of scattered radiation, and other quantities of physical interest.

In this appendix we introduce and define the physical quantities which form the model of multiple scattering of radiation in atmospheres. The classical equation of transfer is derived. In this book, we do not directly solve the transfer equation, which is unstable. Instead, we provide the invariant imbedding treatment of the same problem which leads to much more tractable equations, equations which are solved computationally.

A.1 The Intensity of Radiation

The basic unit of radiative energy transfer is the *intensity*. Consider a three-dimensional coordinate system and an element of surface area centered at a given point $\underline{r}$ normal to the direction $\underline{\Omega}$ of radiation in question. Let $d\Omega$ represent the volume of an elementary solid angle. The amount of radiant energy, dE_ν, in a specified frequency interval $(\nu, \nu + d\nu)$, which traverses an element of area ds during time dt, in the direction $d\varrho$ making an angle θ with the normal to ds is expressed in the form

$$dE_\nu = I_\nu(\underline{r}, \underline{\Omega}, t) \cos\theta d\nu ds d\Omega dt, \tag{A.1}$$

where the quantity I_ν is called the *intensity of radiation*. In other words, the intensity is the quantity of radiant energy in a unit frequency interval passing in a unit of time, a unit of solid angle, and a unit of normal surface area in the given direction of the propagation.

In the limiting process as $d\nu, ds, d\Omega, dt, \rightarrow 0$ in eq. (A.1), the intensity is defined in terms of the position vector $\underline{r}$ of the point, of the direction $\underline{\Omega}$, of the frequency ν, and of the time t. This process also defines a *pencil of radiation*. The factor $\cos\theta$ in eq. (A.1) projects the intensity from normal area to horizontal area.

In the transfer problems of astrophysical, meteorological, biomedical, and other interests, it is usually assumed that the atmospheres (or media) are stratified in plane-parallel layers. If all the optical properties of the medium are constant over the plane, the intensity (in a system of polar coordinates with the z-axis in the direction of a position vector $\underline{r}$) is written in the form

$$I_\nu = I_\nu(z; \theta, \phi; t), \tag{A.2}$$

where z represents the height measured normal to the plane of stratification, θ denotes the polar angle (or colatitude) which a given direction $\underline{\Omega}$ makes with z-axis and ϕ denotes the azimuthal angle of the direction $\underline{\Omega}$ referred to the xz-plane. If I_ν is further independent of ϕ, the radiation field is said to be *axially symmetric* about the z-axis. In eq. (A.2), $\underline{r} = (z)$ and $(\theta, \phi) = \underline{\Omega}$.

In planetary and terrestrial atmospheres we may encounter transfer problems in which the optical properties of the medium depend upon the x- and z-coordinates in a Cartesian system of coordinates, whose z-axis is normal to the plane of stratification. In this case the intensity is expressed

$$I_\nu = I_\nu(x, z; \theta, \phi; t). \tag{A.3}$$

In eq. (A.3), $\underline{r} = (x, z)$ and $\underline{\Omega} = (\theta, \phi)$. Such problems may be called *the transfer problems in a two-dimensional flat layer*.

As an extension of the two-dimensional transfer problem, we can consider the transfer problems in a *three-dimensional flat layer* when the intensity is given respectively by

$$I_\nu = I_\nu(x, y, z; \theta, \phi; t), \tag{A.4}$$

in rectangular coordinates, and

$$I_\nu = I_\nu(z, \rho, \phi; \theta, \psi; t), \tag{A.5}$$

in cylindrical coordinates in which z denotes the geometrical depth normal to the surface of the layer, ρ is the distance from the origin of the coordinates in the xy-plane, θ and ϕ represent respectively the polar angle of the direction $\underline{\Omega}$ and azimuthal angle of the position $\underline{r}$, and ψ is the angle between the projection of the pencil of radiation on the xy-plane and the distance ρ. While in eq. (A.4), $\underline{r} = (x, y, z)$ and $\underline{\Omega} = (\theta, \phi)$, in eq. (A.5), $\underline{r} = (z, \rho, \phi_0)$ and $\underline{\Omega} = (\theta, \psi)$.

In spherical coordinates, the intensity is written in the form

$$I_\nu = I_\nu(r, \theta_0, \phi_0; \theta, \phi; t), \tag{A.6}$$

where r is the radius from the origin of the coordinates, θ_0 (or θ) and ϕ_0 (or ϕ) are respectively the polar and azimuthal angels of the vector $\underline{r}$ (or $\underline{\Omega}$). In eq. (A.6), $\underline{r} = (r, \theta_0, \phi_0)$ and $(\theta, \phi) = \underline{\Omega}$.

The intensity I, defined below as an integral over the whole frequency spectrum, is called the *integrated intensity*:

$$I = \int_0^\infty I_\nu d\nu, \tag{A.7}$$

If the intensity in the medium is independent of the direction Ω at a point, the radiation field is called *isotropic* at the point. Furthermore, if the intensity does not change with the position vector r and the direction Ω, the radiation field is said to be *homogeneous and isotropic*.

A.2 The Absorption and the Scattering Coefficients

In the description of the interaction between radiation and matter, the coefficients of absorption and scattering play an important role. As a result of these types of interaction, the intensity of radiation is weakened by the transformation of radiant energy to thermal energy. The pencil of radiation is in consequence diffusely scattered through the medium until it is absorbed by matter, or escapes from the surface of the medium, if this does not extend to infinity in all directions.

Consider a pencil of radiation, whose intensity is I_ν, traversing a geometrical distance r. As a result of the weakening of radiation by the interactions mentioned above, the intensity I_ν of the pencil suffers a (negative) increment dI_ν. Then, in the absence of emission, we have

$$dI_\nu = -\kappa_\nu \rho I_\nu dr, \tag{A.8}$$

where ρ is the density of the absorbing matter. In the case of true absorption we call κ_ν introduced in this fashion the *monochromatic mass absorption coefficient*. When a beam of radiation traversing the medium suffers energy loss, all or some of the absorbed radiant energy may be scattered from the original direction into other directions. In this case, since the monochromatic mass scattering coefficient σ_ν characterizes the optical properties of the medium, the rate of loss of energy from the incident pencil due to scattering in all directions per unit mass, per unit frequency, and per unit solid angle is given by $-\sigma_\nu I_\nu$. In other words, it is reasonable to assume that, while a part of the reduction in intensity will be observed as scattered radiation, the remaining part will be turned into other forms of energy, e.g., thermal energy. The former represents scattering, while the latter represents true absorption. The sum of the monochromatic mass absorption and monochromatic mass scattering coefficients is sometimes called *the monochromatic mass attenuation coefficient*.

Let the length of radius from the origin of the spherical coordinates in an infinite homogeneous medium, in which a collimated point source of radiation is located, be r. For the attenuation of the beam along the path of length r, dividing both sides of eq. (A.8) by I_ν and integrating with respect to r, we have

$$I_\nu = I_\nu^0 e^{-\tau_\nu}, \tag{A.9}$$

where I_ν^0 is the intensity of the incident beam at the origin and the exponent is

$$\tau_\nu = \int_0^r \kappa_\nu \rho dr. \tag{A.10}$$

The quantity τ_ν, in the flat layer, is called the *optical distance* or optical thickness of the flat layer if r is the geometrical thickness. This measures the reduction in intensity in the absorbing medium, when it is measured normal to the plane of stratification from the boundary.

In accordance with the explanation given above, we introduce the quantity $0 \le \tilde{\omega}_\nu(\tau_\nu) \le 1$ which is a measure of the part of the energy of a given frequency which is scattered directly after an elementary act of scattering with or without redistribution in frequency. Then $1 - \tilde{\omega}_\nu(\tau_\nu)$ is the fraction of the energy which, after an elementary act of scattering, is transformed into other forms of energy and is not reradiated at the same frequency, i.e., it is the part subject to true absorption. The quantity $\tilde{\omega}_\nu(\tau_\nu)$ is usually called the *albedo for single scattering*. The case of $\tilde{\omega}_\nu = 1$ is called the *conservative case* of perfect scattering, because of the conservation of radiant energy.

A.3 The Phase Function

The phase function p (also called non-spherical indicatrix of scattering) describes the scattering of a pencil of radiation without change of frequency from the direction Ω (or θ, ϕ) into the direction Ω' (or θ', ϕ').

The probability that a photon in the direction $\underline{\Omega}$, at the point $\underline{r}$ and at time t will be emitted, after an elementary act of scattering, in the direction interval $(\underline{\Omega}', \underline{\Omega}' + d\underline{\Omega}')$ is expressed in terms of the phase function p as

$$p(\underline{r}, \cos\Theta, t)d\Omega'/4\pi, \tag{A.11}$$

where

$$\cos\Theta = \cos\theta\cos\theta' + \sin\theta\sin\theta'\cos(\phi - \phi'). \tag{A.12}$$

In eq. (A.12) Θ represents the angle between the initial and the final directions, i.e., $\underline{\Omega}$ and $\underline{\Omega}'$. Since the integration of the probability with respect to direction $\underline{\Omega}'$ should be normalized, we have

$$\int_{4\pi} p(\underline{r}, \cos\Theta, t)\frac{d\Omega'}{4\pi} = \frac{1}{4\pi}\int_0^\pi d\theta' \int_0^{2\pi} p(\underline{r}, t; \theta, \phi; \theta', \phi')d\phi' = 1. \tag{A.13}$$

Allowing for the probabilistic measure of the phase function, the positive quantity $p(\underline{r}, \cos\theta, t)$ satisfies the spatial inversion invariance and time-reversal symmetries

$$p(\underline{r}, t; \underline{\Omega}, \underline{\Omega}') = p(\underline{r}, t; \underline{\Omega}', \underline{\Omega}),$$
$$p(\underline{r}, t; \underline{\Omega}, \underline{\Omega}') = p(\underline{r}, t; -\underline{\Omega}', -\underline{\Omega}),$$
$$p(\underline{r}, t; \underline{\Omega}, -\underline{\Omega}') = p(\underline{r}, t; -\underline{\Omega}', \underline{\Omega}) \tag{A.14}$$

where $-\underline{\Omega}$ represents the direction opposite to $\underline{\Omega}$.

If the probability of scattering in any direction is the same, the phase function p is equal to unity and the process is called *isotropic*; otherwise the scattering is *anisotropic*.

A.4 The Emission Coefficient, The Mean Intensity and The Source Function

Assuming that an element of mass dm emits radiation in all directions, the amount of radiant energy at the point $\underline{r}$ inside the solid angle $d\Omega$ in the frequency interval $(\nu, \nu + d\nu)$, in time dt, is given by the expression

$$j_\nu(\underline{r}, \underline{\Omega}, t) d\nu d\Omega dt dm, \tag{A.15}$$

where j_ν is called the *emission coefficient*. In a scattering medium, if both true absorption and scattering are present without change of frequency, the contribution of scattering to the emission coefficient j_ν is given by

$$j_\nu^{(s)}(\underline{r}, \underline{\Omega}, t) = \frac{\sigma_\nu(\underline{r}, t)}{4\pi} \int p(\underline{r}, \cos\Theta; t) I_\nu(\underline{r}, \underline{\Omega}', t) d\Omega', \tag{A.16}$$

where σ_ν is the monochromatic mass scattering coefficient, and $p(\underline{r}, \cos\Theta, t)$ is the phase function defined in Sec. A.3. The integration is over the solid angle.

The case without change of frequency is called *coherent scattering*. In the case of pure scattering, when the scattered radiation is fully re-emitted without change of frequency, we say that we have *monochromatic radiative equilibrium*.

On the other hand, if after an elementary act of scattering a redistribution of energy in frequency takes place according to a certain probability of distribution $p(r, \cos\theta; \nu, \nu')$, where θ is the inclination of the direction of the incident beam to that of the scattered beam, the scattering is called *noncoherent*.

There are two contributions to the emission coefficient $j_\nu(\underline{r}, \underline{\Omega}, t)$, i.e.

$$j_\nu(\underline{r}, \underline{\Omega}, t) = j_\nu^{(s)}(\underline{r}, \underline{\Omega}, t) + j_\nu^{(a)}(\underline{r}, \underline{\Omega}, t). \tag{A.17}$$

where $j_\nu^{(a)} = \kappa_\nu B_\nu(T)$, by Kirchoff's law. And κ_ν as in (A.8) and $B_\nu(T)$ is the Planck function in terms of the local temperature T at $\underline{r}$. In the field of anisotropic radiation the mean value of the intensity weighted by the phase function at a point $\underline{r}$ and at time t is given by

$$j_\nu(\underline{r}, \underline{\Omega}, t) = \frac{1}{4\pi} \int p(\underline{r}, \cos\Theta, t) I_\nu(\underline{r}, \underline{\Omega}', t) d\Omega', \tag{A.18}$$

which represents an amount of scattered radiation which is redistributed inside an elementary cone about the direction. Furthermore, multiplying eq. (A.17) with $\sigma(r, t) d\Omega$, integrating with respect to Ω over the whole range and making use of eq. (A.13), we have $\sigma \int I d\Omega'$ which represents the amount of scattered radiation per unit of mass and of frequency. When the scattering is isotropic, the mean intensity is defined by eq. (A.18) with the unit p–function. In this case the density of radiation is given in terms of the mean intensity (see Sec. 1.5).

The mean value of intensity integrated over the whole spectrum is given by

$$J(\underline{r}, \underline{\Omega}, t) = \int_0^\infty J_\nu'(\underline{r}, \underline{\Omega}, t) d\nu'. \tag{A.19}$$

When true absorption and scattering are both present, the *source function* is given by the ratio of the emission coefficient to the sum of the absorption and scattering coefficients, as follows:

$$S_\nu(\underline{r}, \underline{\Omega}, t) = j_\nu(\underline{r}, \underline{\Omega}, t) / \alpha(\underline{r}, t), \tag{A.20}$$

where $\alpha(\underline{r}, t)$ is the attenuation coefficient.

On using eqs. (A.16) – (A.20) for coherent and anisotropic scattering, we get, generally,

$$S_\nu(\underline{r}, \underline{\Omega}, t) = \frac{\omega_\nu(\underline{r}, t)}{4\pi} \int p(\underline{r}, \cos\Theta, t)$$
$$\cdot I_\nu(\underline{r}, \underline{\Omega}', t) d\Omega' + \frac{j_\nu^{(a)}(\underline{r}, \underline{\Omega}, t)}{\alpha(\underline{r}, t)}. \tag{A.21}$$

The source function gives the rate at which radiation is produced into a unit solid angle per unit volume at r. Here we see that it is due to emission and scattering of radiation previously going in other directions $\underline{\Omega}'$.

A.5 The Net Flux and the Density of Radiation

The net flow of radiant energy in all directions, dE_ν, in a specified frequency interval $(\nu, \nu + d\nu)$, passing through an element of area ds normal to the unit vector $\underline{\Omega}_0$, whose direction coincides with that of the position vector $\underline{r}$ of the point P under consideration, within an elementary solid angle $d\Omega'$, during time dt, is given by

$$dE_\nu = d\nu ds dt \int I_\nu(\underline{r}, \underline{\Omega}', t) \cos\Theta d\Omega', \tag{A.22}$$

where Θ is the angle between the direction $\underline{\Omega}'$ of the pencil of scattered radiation and the direction $\underline{\Omega}_0$, and the integration is carried out over all solid angels.

Then, as a rate of flow of radiant energy traversing an elementary area ds per unit area, per unit frequency interval, and per unit time, the net flux is defined by

$$\pi F_\nu = \int I_\nu(\underline{r}, \underline{\Omega}', t) \cos\Theta d\Omega' \tag{A.23}$$

where, taking the direction cosines of the unit vectors $\underline{\Omega}_0$ and $\underline{\Omega}'$ respectively to be (ℓ_0, m_0, n_0) and (ℓ', m', n') for the Cartesian frame of reference, we have

$$\cos\Theta = \ell_0\ell' + m_0 m' + n_0 n'. \tag{A.24}$$

Alternatively,

$$F_{\nu;\ell_0,m_0,n_0} = \ell_0 F_{\nu,x} + m_0 F_{\nu,y} + n_0 F_{\nu,z}, \tag{A.25}$$

where

$$F_{\nu,x} = \int I_\nu \ell' d\Omega'/\pi,$$

$$F_{\nu,y} = \int I_\nu m' d\Omega'/\pi,$$

$$F_{\nu,z} = \int I_\nu n' d\Omega'/\pi. \tag{A.26}$$

In the foregoing, $F_{\nu,x'}$, $F_{\nu,y'}$, and $F_{\nu,z}$ represent respectively the fluxes crossing elementary areas normal to the x-, y- and z-axes.

For a radiation field in which the z-axis is in the direction of Ω_0, the net flux F_ν is given by

$$\pi F_\nu = \int_0^{2\pi} d\phi' \int_0^\pi I_\nu(z; \theta', \phi'; t) \sin\theta' \cos\theta' d\theta', \tag{A.27}$$

since $\Theta = \theta'$ and $d\Omega' = \sin\theta' d\theta' d\phi'$.

Allowing for the net flow, F_ν may be rewritten in the form

$$F_\nu = F_\nu^+ + F_\nu^-, \tag{A.28}$$

where

$$F_\nu^+ = \int_0^{\pi/2} d\theta' \int_0^{2\pi} I_\nu \sin\theta' \cos\theta' d\phi',$$

$$F_\nu^- = \int_{\pi/2}^{\pi} d\theta' \int_0^{2\pi} I_\nu \sin\theta' \cos\theta' d\phi'. \tag{A.29}$$

Here F_ν^+ and F_ν^- represent respectively the flux of outward radiation and that of inward radiation.

For an axially symmetric radiation field in which the intensity is independent of the azimuthal angle ϕ, the net flux F_ν reduces to

$$F_\nu(z,t) = 2\int_0^{\pi} I_\nu(z,\theta',t) \sin\theta' \cos\theta' d\theta'. \tag{A.30}$$

Furthermore, the integrated net flux over the whole frequency spectrum is

$$F(z,t) = \int_0^\infty F_\nu(z,t)d\nu. \tag{A.31}$$

If there is no distribution of heat sources in the medium, according to the principle of energy conservation, the net flux of radiation integrated over the whole spectrum must be constant. This state of affairs is usually called *radiative equilibrium*. Physically speaking, when radiation is the sole agency of energy transfer, the total radiation absorbed by an element of mass is equal to the total emission by the same element.

On the other hand (allowing for the fact that the quantity F is an average of the emergent intensity), in local thermodynamical equilibrium the integrated intensity is given in terms of the local temperature T by the relation $\sigma T^4/\pi$.

The amount of radiant energy per unit volume, within the specified frequency interval $(\nu, \nu + d\nu)$, in a unit time, passing through an elementary volume dv which contains a given point P is called the *density of radiation* $u_\nu d\nu$.

Consider a cavity of an infinitesimal volume v, the inner surface of whose elementary area emits radiation. Upon recalling eq. (A.1), the amount of radiant energy emitted by ds, within an elementary solid angle $d\Omega$ per unit time is given by

$$dE_\nu = I_\nu(\underline{r}, \underline{\Omega}, t)d\nu ds \cos\theta d\Omega, \tag{A.32}$$

where

$$d\Omega = \frac{\cos\theta' ds'}{r^2}. \tag{A.33}$$

In eq. (A.33), ds' represents an elementary area of the internal surface of the cavity at which the specific pencil of radiation in the direction Ω intersects, θ' is the angle between the line joining the two elements and the normal to ds', and r is the distance between ds and ds'.

The radiation energy contained in volume v is given by

$$vu_\nu = \int\int I_\nu(\underline{r},\underline{\Omega},t)\cos\theta ds d\Omega\frac{\ell}{c}, \tag{A.34}$$

where ℓ and c are respectively the length traversed by the pencil of radiation under consideration, and the speed of light. From the relation $v = \ell\cos\theta ds$, we have

$$u_\nu(\underline{r},t) = \frac{\ell}{c}\int I_\nu(\underline{r},\underline{\Omega},t)d\Omega. \tag{A.35}$$

Hence, recalling the mean intensity for isotropic scattering defined in (A.19), eq. (A.35) becomes

$$u_\nu(\underline{r},t) = \frac{4\pi}{c}J_\nu(\underline{r},t). \tag{A.36}$$

The density of integrated radiation over the whole spectrum is

$$u(\underline{r},t) = \int_0^\infty u_\nu(\underline{r},t)d\nu. \tag{A.37}$$

A.6 The Equation of Transfer

The difference between the intensity of radiation with specified direction Ω and frequency ν leaving the infinitesimal right cylinder of height cdt in the direction Ω and with base area ds centered at $\underline{r}$, during a time dt, is given by

$$dE_\nu = [I_\nu(\underline{r},\underline{\Omega}+c\underline{\Omega}dt,t+dt) - I_\nu(\underline{r},\underline{\Omega},t)]d\nu ds d\Omega dt, \tag{A.38}$$

where c is the speed of light. Expanding the expression in square brackets in (A.38) in a Taylor series and keeping only the linear term, we have the net amount of the change of intensity dE_ν,

$$dE_\nu = [\frac{\partial I_\nu}{\partial t} + c\underline{\Omega}\nabla I_\nu(\underline{r},\underline{\Omega},t)]d\nu ds d\Omega dt^2. \tag{A.39}$$

The directional derivative of the intensity in the direction $\underline{\Omega}$ expressed in terms of the three-dimensional orthogonal curvilinear coordinates is

$$\underline{\Omega}\nabla I_\nu = \underline{\Omega} \sum_i \frac{a_i}{h_i}\frac{\partial I_\nu}{\partial \xi_i} = \sum_i \frac{\Omega_i}{h_i}\frac{\partial I_\nu}{\partial \xi_i}, \tag{A.40}$$

or, alternatively,

$$\underline{\Omega}\nabla I_\nu = \frac{dI_\nu}{d\ell} = \sum_i \Omega_i \frac{\partial I_\nu}{\partial \xi_i}, \tag{A.41}$$

where ℓ is a distance laid off along Ω.

In eq. (A.40), the $a_i's$ denote unit vectors, i.e., each tangent to the corresponding coordinates line of the curvilinear system which goes through the point, and the $h_i's$ are scale factors for the coordinates ξ_i. In general, a_i and h_i vary from point to point in space. Since a specified direction $\underline{\Omega}$ is a unit vector, the directional derivative of the intensity in terms of the scalar product of the gradient of intensity with the vector $\underline{\Omega}$ represents the rate of change of the intensity in the direction $\underline{\Omega}'$, a true scalar invariant.

The formulation of the equation of transfer for coherent and anisotropic scattering can thus be made, allowing for the difference between the flow of radiant energy into and out of the infinitesimal right cylinder, i.e., the difference between absorption and scattering of radiation inside the cylinder. The change in the intensity will be due to

(i) the attenuation of the intensity of the specified beam

$$dA_\nu = -\alpha_\nu(\underline{r},t)I_\nu(\underline{r},\underline{\Omega},t)cd\nu dsd\Omega dt^2, \tag{A.42}$$

where $\alpha_\nu(\underline{r},t)$ is the monochromatic volume attenuation coefficient,

(ii) scattering of radiation of direction $\underline{\Omega}'$ into the given direction $\underline{\Omega}$,

$$dR_\nu = \frac{c_\nu(\underline{r},t)}{4\pi} \int p(\underline{r},t;\underline{\Omega},\underline{\Omega}')I_\nu(\underline{r},\underline{\Omega}',t)d\Omega'cd\nu dsd\Omega dt^2, \tag{A.43}$$

where p is the phase function, and

(iii) emission due to the source, if any, of photons of the given frequency and direction within the cylinder

$$dN_\nu = \alpha_\nu(\underline{r},t)S_\nu(\underline{r},\underline{\Omega},t)cd\nu dsd\Omega dt^2, \tag{A.44}$$

where $S_\nu(\underline{r},\underline{\Omega},t)$ represents the source density (or the distribution of energy sources) per unit volume, per unit time, per unit steradian and per unit frequency.

On equating the total change of the intensity as given by (A.41) to the sum of these individual contributing factors (A.42) − (A.44), and dividing by the common factor $cd\nu dsd\Omega dt^2$, the time-dependent *equation of transfer* in three-dimensional medium with coherent and anisotropic scattering may be written in the form

$$\frac{1}{c}\frac{dI_\nu(\underline{r},\underline{\Omega},t)}{dt} + \alpha_\nu(\underline{r},t)I_\nu(\underline{r},\underline{\Omega},t) = \alpha_\nu(\underline{r},t)\Sigma_\nu(\underline{r},\underline{\Omega},t), \qquad (A.45)$$

where

$$\frac{1}{c}\frac{dI_\nu}{dt} = \frac{1}{c}\frac{\partial I_\nu}{\partial t} + \underline{\Omega}^- I_\nu, \qquad (A.46)$$

and the source function $\Sigma_\nu(\underline{r},\underline{\Omega},t)$ is

$$\Sigma_\nu(\underline{r},\underline{\Omega},t) = \frac{\omega_\nu(\underline{r},t)}{4\pi}\int p(\underline{r},t;\underline{\Omega},\underline{\Omega}')I_\nu(\underline{r},\underline{\Omega}',t)d\Omega'$$
$$+ S_\nu(\underline{r},\underline{\Omega},t). \qquad (A.47)$$

For the noncoherently scattering atmosphere the time-dependent equation of transfer is given by eq. (A.45) whose source function is

$$\Sigma_\nu(\underline{r},\underline{\Omega},t) = \frac{\omega_\nu(\underline{r},t)}{4\pi\sigma(\underline{r},t)}\int_0^\infty \sigma'_\nu(\underline{r},t)d\nu'$$
$$\cdot \int I'_\nu(\underline{r},\underline{\Omega}',t)d\Omega' + S_\nu(\underline{r},\underline{\Omega},t). \qquad (A.48)$$

The following are some simple unit source densities.

(a) Denoting, as usual, the Dirac delta function by δ, we have for collimated point source,

$$S_\nu(\underline{r},\underline{\Omega},t) = \delta(\underline{r} - \underline{r}_0)\delta(\underline{\Omega} - \underline{\Omega}_0)\delta(\nu - \nu_0)\delta(t - t_0), \qquad (A.49)$$

(b) For a simple isotropic point source,

$$S_\nu(\underline{r},\underline{\Omega},t) = \frac{1}{4\pi(\underline{r} - \underline{r}_0)}\delta(\underline{r} - \underline{r}_0)\frac{1}{4\pi}$$
$$\cdot \delta(\nu - \nu_0)\delta(t - t_0), \qquad (A.50)$$

(c) For the collimated plane source in which the plane passes through the point $\underline{r}_0$ and is normal to the unit vector $\underline{m}$,

$$S_\nu(\underline{r},\underline{\Omega},t) = \delta(\underline{r}\cdot\underline{m} - \underline{r}_0\cdot\underline{m})\delta(\underline{\Omega} - \underline{\Omega}_0)$$
$$\cdot \delta(\nu - \nu_0)\delta(t - t_0), \qquad (A.51)$$

(d) For an isotropic plane source, where z is normal to the plane,

$$S_\nu(\underline{r},\underline{\Omega},t) = \delta(z - z_0)\frac{1}{4\pi}\delta(\nu - \nu_0)\delta(t - t_0), \qquad (A.52)$$

In eqs. (A.49) – (A.52), in place of $\delta(t-t_0)$, the unit step function $H(t-t_0)$ may be used as follows:

$$H(t - t_0) = \begin{cases} 0 \text{ for } t < t_0 \\ 1 \text{ for } t > t_0 \end{cases} \tag{A.53}$$

It is of interest to mention that the collimated point source density can be expressed in terms of the superposition of the collimated plane source density, since Fourier expansion of the Dirac delta function yields the relation

$$\delta(\underline{r} - \underline{r}_0) = \frac{1}{(2\pi)^3} \int d\underline{p} \int_{-\infty}^{\infty} e^{-ips} \delta(\underline{r} \cdot \underline{m} - \underline{r}_0 \cdot \underline{m} - s) ds, \tag{A.54}$$

where $\underline{m}$ denotes the direction of the vector $\underline{p} = p\underline{m}$.

Furthermore, since the transfer eq. (A.46) for noncoherent scattering is linear, the solution of eq. (A.47) with the source function (A.48) for any source density $S_\nu(\underline{r}, \underline{\Omega}, t)$ is expressed in terms of the solution of eq. (A.45) with eq. (A.46) and collimated point source density (A.49),

$$I_\nu(\underline{r}, \underline{\Omega}, t) = \int I_p(\nu, \underline{r}, \underline{\Omega}, t; \nu', \underline{r}', \underline{\Omega}', t')$$
$$\cdot S_\nu'(\underline{r}', \underline{\Omega}', t') d\nu' d\underline{r}' d\underline{\Omega}' dt', \tag{A.55}$$

where $I_p(\nu, \underline{r}, \underline{\Omega}, t; \nu_0, \underline{r}_0, \underline{\Omega}_0, t_0)$ is the intensity of a pencil of radiation at the point $\underline{r}$ in the direction $\underline{\Omega}$ at time t from a point source center at the point $\underline{r}_0$ and emitting radiation in the direction $\underline{\Omega}_0$ at time t_0.

Setting, in eq. (A.45),

$$LI_p(\nu, \underline{r}, \underline{\Omega}, t; \nu_0, \underline{r}_0, \underline{\Omega}_0, t_0) =$$
$$\delta(\nu - \nu_0)\delta(\underline{r} - \underline{r}_0)\delta(\underline{\Omega} - \underline{\Omega}_0)\delta(\nu - \nu_0)\delta(t - t_0)\alpha_\nu(\underline{r}, t), \tag{A.56}$$

where the integro-differential operator L is to be

$$L = \frac{1}{c}\frac{\partial}{\partial t} + \underline{\Omega} \cdot \nabla$$
$$+ \alpha_\nu(\underline{r}, t)\{1 - \frac{\omega_\nu(\underline{r}, t)}{4\pi} \int p(\underline{r}, t; \underline{\Omega}, \underline{\Omega}') d\underline{\Omega}'\}, \tag{A.57}$$

and assuming eq. (A.55), we obtain the operation of L on the intensity $I_\nu(\underline{r}, \underline{\Omega}, t)$

$$LI_\nu = \int LI_p(\nu, \underline{r}, \underline{\Omega}, t; \nu', \underline{r}', \underline{\Omega}', t') S'_\nu(\underline{r}', \underline{\Omega}', t') d\nu' d\underline{r}', d\Omega', dt'$$

$$= \int \delta(\nu - \nu_0)\delta(\underline{r} - \underline{r}')\delta(\underline{\Omega} - \underline{\Omega}')\delta(t - t')\alpha_\nu(\underline{r}, t)$$

$$\cdot S'_\nu(\underline{r}', \underline{\Omega}', t') d\nu' d\underline{r}', d\Omega', dt'$$

$$= S_\nu(\underline{r}, \underline{\Omega}, t)\alpha_\nu(\underline{r}, t). \tag{A.58}$$

This is eq. (A.45).

When we consider Cartesian coordinates as a standard frame of reference, the relationship between h_i and ξ_i is given by

$$h_i^2 = \left(\frac{\partial x}{\partial \xi_i}\right)^2 + \left(\frac{\partial y}{\partial \xi_i}\right)^2 + \left(\frac{\partial z}{\partial \xi_i}\right)^2. \tag{A.59}$$

Hence, for the three-dimensional medium the expression for $\Omega \cdot \nabla I_\nu$ becomes as below:

(i) For Cartesian coordinates $\xi_1 = x$, $\xi_2 = y$, $\xi_3 = z$, $h_1 = h_2 = h_3 = 1$. Then

$$\Omega \cdot \nabla I_\nu(x, y, z; \theta, \phi; t) = \Omega_x^* \frac{\partial I_\nu}{\partial x} + \Omega_y^* \frac{\partial I_\nu}{\partial y} + \Omega_z^* \frac{\partial I_\nu}{\partial z}, \tag{A.60}$$

where θ and ϕ are respectively the colatitude and the azimuth of the unit vector $\underline{\Omega}$ (see Fig. 1) and where

$$\Omega_x^* = \frac{\partial x}{\partial \ell} = \sin\theta \cos\phi,$$

$$\Omega_y^* = \frac{\partial y}{\partial \ell} = \sin\theta \cos\phi,$$

$$\Omega_z^* = \frac{\partial z}{\partial \ell} = \cos\theta. \tag{A.61}$$

(ii) For spherical coordinates $\xi_1 = r$, $\xi_2 = \theta_0$, $\xi_3 = \phi$, $h_1 = 1,$, $h_2 = r$, $h_3 = r\sin\theta_0$, then

$$\Omega \cdot \nabla I_\nu(r, \theta_0, \phi_0; \theta, \phi, t) = \Omega_r^* \frac{\partial I_\nu}{\partial r} + \Omega_{\theta_0}^* \frac{\partial I_\nu}{\partial \theta_0} + \Omega_{\phi_0}^* \frac{\partial I_\nu}{\partial \phi_0} + \tag{A.62}$$

where the position vector $\underline{r}$ and the unit vector Ω are given respectively by (r, θ_0, ϕ_0) and (θ, ϕ) in which θ_0(or θ) and ϕ_0(or ϕ) represent the colatitude and the azimuth of the vector $\underline{r}$(or Ω), respectively, and

$$\Omega_r^* = \frac{\partial r}{\partial \ell} = \cos\theta \cos\theta_0 + \sin\theta \sin\theta_0 \cos(\phi - \phi_0),$$

$$\Omega_{\theta_0}^* = \frac{\partial \theta_0}{\partial \ell} = -\frac{1}{r}\{\cos\theta \sin\theta_0 - \sin\theta \cos\theta_0 \cos(\phi - \phi_0)\},$$

$$\Omega_{\phi_0}^* \frac{\partial \phi_0}{\partial \ell} = \frac{1}{r} \frac{\sin\theta}{\sin\theta_0} \sin(\phi - \phi_0). \qquad (A.63)$$

(iii) For cylindrical coordinates $\xi_1 = z$, $\xi_2 = \rho$, $\xi_3 = \phi_0$, $h_1 = 1$, $h_2 = 1$, $h_3 = \rho$ then,

$$\underline{\Omega} \cdot \nabla I_\nu (z, \rho, \phi_0; \theta, \psi; t) = \Omega_z^* \frac{\partial I_\nu}{\partial z} + \Omega_\rho^* \frac{\partial I_\nu}{\partial \rho} + \Omega_{\phi_0}^* \frac{\partial I_\nu}{\partial \phi_0}, \qquad (A.64)$$

where the position vector $\underline{r}$ and the unit vector $\underline{\Omega}$ are given respectively in terms of (z, ρ, ϕ_0) and of (θ, ψ) $(\psi = \phi - \phi_0)$, where θ is the colatitude of the unit vector $\underline{\Omega}$ and ϕ(or ϕ_0) is the azimuth of the unit vector $\underline{\Omega}$(or $\underline{r}$),

$$\Omega_z^* = \frac{\partial z}{\partial \ell} = \cos\theta,$$

$$\Omega_\rho^* = \frac{\partial \rho}{\partial \ell} = \sin\theta \cos\psi,$$

$$\Omega_{\phi_0}^* \frac{\partial \phi_0}{\partial \ell} = \frac{1}{\rho} \sin\theta \sin\psi. \qquad (A.65)$$

If the intensity is symmetric about the z–axis, i.e., a function of $\psi = \phi - \phi_0$, we have

$$\underline{\Omega} \cdot \nabla I_\nu (z, \rho, \theta, \psi, t) = \Omega_z^* \frac{\partial I_\nu}{\partial z} + \Omega_\rho^* \frac{\partial I_\nu}{\partial \rho} + \Omega_\psi^* \frac{\partial I_\nu}{\partial \psi}, \qquad (A.66)$$

where Ω_z^*, Ω_ρ^* and Ω_ψ^* are equal respectively to $\Omega_z^*, \Omega_\rho^*$, and $-\Omega_{\phi_0}^*$ given in eq. (A.65).

For the two-dimensional medium, we obtain the following equations:
(i) In Cartesian coordinates

$$\Omega \cdot \nabla I_\nu (x, z; \theta, \phi; t) = \Omega_x^* \frac{\partial I_\nu}{\partial x} + \Omega_z^* \frac{\partial I_\nu}{\partial z}. \qquad (A.67)$$

(ii) In spherical coordinates

$$\Omega \cdot \nabla I_\nu (r, \theta_0; \theta, \phi; t) = \Omega_r^* \frac{\partial I_\nu}{\partial r} + \Omega_{\theta_0}^* \frac{\partial I_\nu}{\partial \theta_0}, \qquad (A.68)$$

where Ω_r^* and $\Omega_{\theta_0}^*$ are given by eq. (A.63).
(iii) In cylindrical coordinates

$$\Omega \cdot \nabla I_\nu(z, \rho, \theta, \psi, t) = \Omega_z^* \frac{\partial I_\nu}{\partial z} + \Omega_\rho^* \frac{\partial I_\nu}{\partial \rho}, \tag{A.69}$$

where $\psi = \phi - \phi_0$, and Ω_z^* and Ω_ρ^* are given by eq. (A.65). Eq. (A.67) reduces to eq. (A.67) when $\phi_0 = 0$.

When the system is invariant with respect to reflection in a plane, reflection through a point and rotation about the z-axis, respectively, the system is said to have plane, spherical and cylindrical *symmetry*, respectively. Furthermore, if the angular distribution of the emergent radiation has plane, spherical and cylindrical symmetry, respectively, then the description space is said to be that of plane, spherical and cylindrical *geometry*.

(i) In a one-dimensional medium the expression for $\Omega \cdot \nabla I_\nu$ simplifies and becomes:

$$\underline{\Omega} \cdot \nabla I_\nu(z; \theta; t) = \cos \theta \frac{\partial I_\nu}{\partial z} \tag{A.70}$$

for plane geometry, where the z-axis is normal to the plane of stratification, and θ is the colatitude of the unit vector $\underline{\Omega}$.

(ii) For spherical geometry, where r is the distance from the origin of spherical symmetry, and ψ is the angle between the direction $\underline{\Omega}$ and the radius r.

$$\Omega \cdot \nabla I_\nu(r; \psi; t) = \cos \psi \frac{\partial I_\nu}{\partial r} - \frac{\sin \psi}{r} \frac{\partial I_\nu}{\partial \psi} \tag{A.71}$$

(iii) For cylindrical geometry

$$\Omega \cdot \nabla I_\nu(\rho; \theta, , \psi; t) = \sin \theta \cos \psi \frac{\partial I_\nu}{\partial \rho} - \frac{\sin \theta \sin \psi}{\rho} \frac{\partial I_\nu}{\partial \psi}, \tag{A.72}$$

where ρ is the distance measured perpendicularly from the z–axis of the infinitely long cylinder, θ the colatitude of $\underline{\Omega}$, and ψ the angle between the projection of $\underline{\Omega}$ in the xy-plane and the distance ρ, i.e., $\psi = \phi - \phi_0$.

A.7 Discussion

The equation of transfer for a given situation must be solved subject to the appropriate boundary conditions. Thus the formulation leads to a two-point boundary-value problem. While the differential equations are linear, they are unstable – which makes numerical solution impractical.

We have defined the fundamental physical quantities used in formulating mathematical models for multiple scattering problems. They have been incorporated in the equations of transfer, just presented, for various cases of interest. For references see [1]–[6].

References

1. S. Chandrasekhar, *Radiative Transfer*, Dover, New York, 1960.

2. V. V. Sobolev, *Dokl. Akad. Nauk SSSR*, Vol. 179, 1968, p. 41 [*Sov. Phys.-Dokl.*, Vol. 13, 1968, p. 180].

3. E. A. Milne, *Thermodynamics of the Stars*, Handbuck der Astrophysik, Band iii/1, pp. 65–173, Springer, Berlin, 1930.

4. V. A. Ambarzumian, *Diffuse Reflection of Light by a Foggy Medium*, Compt. Rendus Acad. Sci. U.S.S.R., Vol. 38, 1943, pp. 229–232.

5. I. W. Busbridge, *The Mathematics of Radiative Transfer*, Cambridge University Press, Cambridge, 1960.

6. V. Kourganoff, *Basic Methods in Transfer Problems*, Dover Publications, New York, 1963.

B. Appendix B
Derivation and Validation of Imbedding Equations

We consider anew the basic functions of radiative transfer and the derivation of their invariant imbedding Cauchy equations. We do this, not physically as with the addition of a thin layer, but by first deriving from the physics alternate mathematical models, and from them we analytically derive the imbedding equations. The functions we treat in this Appendix are the source function, the reflection function, and the internal intensity functions. Other functions of interest may also be studied in this way.

B.1 Introduction

For the source function, we derive the integral equation which it satisfies. From that equation we derive the imbedding equations. For the reflection function, we have its definition in terms of the source function, and we obtain its imbedding equations. For the internal intensity functions, we also have their definitions in terms of the source function, and we derive their Cauchy equations. The internal intensity functions also satisfy a two-point boundary value problem, so it is significant that the solution of such a boundary value problem also satisfies a Cauchy problem.

The conclusion we draw from the discussion of this Appendix and other previous work is that integral equations, two-point boundary value problems, and Cauchy problems of radiative transfer are equivalent. All are valid mathematical formulations of radiative transfer models. Some have greater utility theoretically, while others are useful computationally.

B.2 Source Function

B.2.1 Integral equation

Starting from the definition of the source function, we derive the integral equation which it satisfies. Then, via purely mathematical operations, we obtain the Cauchy system. This system is exactly the same as that which is derived using physical imbedding principles, and thus validates the physical method.

Consider an inhomogeneous plane-parallel isotropically scattering medium of optical thickness, x, whose albedo for single scattering as a function of optical altitude t is $\lambda(t)$, $0 \leq t \leq x$. The source function is defined such that $J(t, x, u)dt$ is the production of scattered radiation in a cylinder of unit cross-sectional area between altitude t and $t + dt$ in the slab of thickness x per unit solid angle per unit time, due to uniform parallel radiation of net flux π per unit normal area incident at the top with cosine u.

The source function will be shown to satisfy what is called the Milne integral equation, which is a Fredholm integral equation of the second kind [1, 2]. The equation is

$$J(t, x, u) = \frac{1}{4}\lambda(t)e^{-(x-t)/u} + \frac{1}{2}\lambda(t)\int_0^x J(t', x, u)E_1(|t - t'|)dt', \quad \text{(B.1)}$$

for $0 \leq t \leq x$; x and u being parameters, where $E_1(\cdot)$ is the first exponential integral [3],

$$E_1(r) = \int_0^1 e^{-r/v}\frac{dv}{v}, \quad r > 0. \quad \text{(B.2)}$$

The first exponential integral function has a physical meaning in radiative transfer:

$E_1(r)dt = $ the fraction of the radiation starting at a certain point that is absorbed in a small dt interval at a distance r away (with all polar angles in the upward or downward directions).

In the Fredholm integral equation for the source function $J(t, x, u)$, consider J to be a function of t. The function J appears on the left hand side and again under the integral on the right hand side. The forcing function is the first term on the right hand side, while the kernel of the integral equation is the coefficient of J in the second term. An equivalent form of the Milne equation, using the definition of the first exponential integral is

$$J(t, x, u) = \frac{1}{4}\lambda(t)e^{-(x-t)/u} + \frac{1}{2}\lambda(t)\int_0^x J(t', x, u)\int_0^1 e^{-|t-t'|/v}\frac{dv}{v}dt'. \quad \text{(B.3)}$$

A Fredholm integral equation of the second kind has the form,

$$y(t) = f(t) + \int_0^x k(t, t')y(t')dt', \quad \text{(B.4)}$$

in which $f(t)$ is the forcing function and $k(t, t')$ is the kernel, and $y(t)$ is the unknown function.

Since Fredholm integral equations of the second kind are linear in the forcing function, a superposition principle can be used to represent the solution of a similar equation, i.e., with the same kernel but with a linear combination of forcing functions, in terms of a linear combination of solutions. By example, consider the two integral equations,

$$y_1(t) \;=\; f_1(t) + \int_0^x k(t,t')y_1(t')dt',$$

$$y_2(t) \;=\; f_2(t) + \int_0^x k(t,t')y_2(t')dt'.$$

If $y(t)$ satisfies an integral equation with the same kernel and a forcing function which is a linear combination of the two other forcing functions, i.e.,

$$y(t) = f(t) + \int_0^x k(t,t')y(t')dt',$$

where

$$f(t) = f_1(t) + f_2(t),$$

then the solution is

$$y(t) = y_1(t) + y_2(t).$$

B.2.2 Derivation of the integral equation

We consider the quantity of radiation $J(t,x,u)dt$ and express it as the sum of two terms. The first term is

$$\frac{1}{4}\lambda(t)e^{-(x-t)/u}dt.$$

This term arises by considering the following:

$\pi =$ radiation per unit normal area incident at the top

$v =$ component of radiation per unit normal area

$e^{-(x-t)/u} =$ fraction of radiation at x that reaches altitude t

$dt/v =$ fraction of radiation going with direction cosine v that is absorbed between $t + dt$ and t

$\frac{1}{4\pi}\lambda(t) =$ fraction of radiation at t that is scattered isotropically into a unit solid angle.

Thus we have the forcing function,

$$\pi v e^{-(x-t)/u}\frac{dt}{v}\left(\frac{1}{4\pi}\right)\lambda(t) = \frac{1}{4}\lambda(t)e^{-(x-t)/u}dt.$$

The second term arises from scatterings at other levels and may be written, when the integral over altitude is broken into two parts:

$$\frac{1}{2}\lambda(t)\int_0^t J(t',x,u)\int_0^1 e^{-(t-t')/v}\frac{dv}{v}dt'$$

$$+\frac{1}{2}\lambda(t)\int_t^x J(t',x,u)\int_0^1 e^{-(t'-t)/v}\frac{dv}{v}dt'.$$

The rate of production of scattered radiation at level t' is $J(t', x, u)$. For the first expression, $t > t'$, the scattered radiation with direction cosine v travels upward directly to level t. The factors are

$$e^{-(t-t')/v} = \text{fraction of radiation at } t' \text{ that reaches altitude } t,$$

$$\frac{dt}{v} = \text{fraction of radiation absorbed between } t \text{ and } t + dt$$

$$\frac{1}{4\pi}\lambda(t) = \text{fraction of radiation scattered into a unit solid angle.}$$

Thus, we have the contribution,

$$2\pi e^{-(t-t')/v}\frac{dt}{v}\left(\frac{1}{4\pi}\right)\lambda(t)J(t', x, u), \quad t > t',$$

which must be integrated over direction cosines between 0 and 1, and over altitudes from 0 to t. This leads to the expression,

$$\int_0^t \frac{1}{2}\lambda(t)\int_0^1 e^{-(t-t')/v}\frac{dv}{v}J(t', x, u)dt'dt.$$

For radiation that travels downward from $t' > t$ to altitude t, we have the contribution,

$$\int_t^x \frac{1}{2}\lambda(t)\int_0^1 e^{-(t'-t)/v}\frac{dv}{v}J(t', x, u)dt'dt.$$

Combining the last two expressions and dividing by dt, we have the integral term in the integral equations of (B.1) and (B.3),

$$\frac{1}{2}\lambda(t)\int_0^x J(t', x, u)\int_0^1 e^{-|t-t'|/v}\frac{dv}{v}dt' = \frac{1}{2}\lambda(t)\int_0^x J(t', x, u)E_1(|t - t'|)dt'.$$

Therefore, the integral eq. (B.1) or (B.3) follows.

Solution of the Fredholm integral equation may appear to be straightforward using quadrature formulas to replace the double integral and solving a system of linear algebraic equations for a set of discrete source functions. This is ill-advised, since the Milne equation is notoriously ill-conditioned, as many have learned to their chagrin. Instead, the Cauchy method is recommended. We proceed to derive the Cauchy problem from the Milne integral equation.

B.2.3 The Φ function

Let us introduce the Sobolev function $\Phi(t, x)$ as the solution of the Milne integral equation with forcing function

$$\frac{1}{2}\lambda(t)E_1(x - t) = \frac{1}{2}\lambda(t)\int_0^1 e^{-(x-t)/v}\frac{dv}{v}$$

$$= 2\int_0^1 \left\{\frac{1}{4}\lambda(t)e^{-(x-t)/v}\frac{dv}{v}\right\}.$$

From the superposition principle of Fredholm integral equations, we have the solution $\Phi(t, x)$ expressed in terms of the solution of the integral equation with the forcing function in the curly brackets, i.e.,

$$\Phi(t, x) = 2 \int_0^1 J(t, x, v) \frac{dv}{v}. \tag{B.5}$$

From the nature of the forcing function we see that the function $\Phi(t, x)$ is the source function with omnidirectional illumination of net flux π per unit normal area, a very important function in the theory of radiative transfer [4].

B.2.4 Analytical derivation of the imbedding equations

Previously, we derived the imbedding equations for J using a physical imbedding approach. Here we obtain the same equations analytically. Begin with the integral Eq. (B.1), in which $J(t, x, u)$ is now regarded as a function of x as well as of t, u being held fixed. Differentiate on both sides of Eq. (B.1) with respect to x. We designate the partial derivative, on the left hand side, with respect to x as $J_x(t, x, u)$. On the right, differentiation of the first term yields

$$-\frac{1}{u}\frac{1}{4}\lambda(t)e^{-(x-t)/u}.$$

The integral term gives rise to two terms, when differentiated, because x appears in two places. The two terms are

$$\frac{1}{2}\lambda(t)\left\{ J(x, x, u)E_1(x - t) + \int_0^x J_x(t', x, u)E_1(|t - t'|/v)dt' \right\}.$$

The equation for $J_x(t, x, u)$ is an integral equation for $J_x(t, x, u)$:

$$\begin{aligned} J_x(t, x, u) \quad = \quad & -\frac{1}{u}\frac{1}{4}\lambda(t)e^{-(x-t)/u} \\ & +\frac{1}{2}\lambda(t)J(x, x, u)E_1(x - t) \\ & +\frac{1}{2}\lambda(t)\int_0^x J_x(t', x, u)E_1(|t - t'|/v)dt'. \end{aligned} \tag{B.6}$$

It has the same kernel as the previous integral equations. From superposition, we obtain the solution of integral Eq. (B.6) above as the linear combination of solutions of the integral equations for J and for Φ,

$$J_x(t, x, u) = -\frac{1}{u}J(t, x, u) + J(x, x, u)\Phi(t, x), \quad x > t. \tag{B.7}$$

This can also be written as

$$J_x(t, x, u) = -\frac{1}{u}J(t, x, u) + 2J(x, x, u)\int_0^1 J(t, x, v)\frac{dv}{v}, \quad x > t. \tag{B.8}$$

This is one of the desired invariant imbedding Cauchy equations.

To obtain the initial condition on J, we set $t = x$ in the integral equation (B.1) and we have

$$
\begin{aligned}
J(x, x, u) &= \frac{1}{4}\lambda(x) + \frac{1}{2}\lambda(x) \int_0^x J(t', x, u) E_1(x - t') dt' \qquad\qquad \text{(B.9)} \\
&= \frac{1}{4}\lambda(x) + \frac{1}{2}\lambda(x) \int_0^x J(t', x, u) \int_0^1 e^{-(x-t')/v} \frac{dv}{v} dt' \\
&= \frac{1}{4}\lambda(x) + \frac{1}{2}\lambda(x) \int_0^1 \frac{dv}{v} \int_0^x J(t', x, u) e^{-(x-t')/v} dt' \\
&= \frac{1}{4}\lambda(x) \left\{ 1 + 2 \int_0^1 r(x, v, u) dv \right\}.
\end{aligned}
$$

In Eq. (B.9) we have introduced the reflection function r,

$$
r(x, v, u) = \frac{1}{v} \int_0^x J(t', x, u) e^{-(x-t')/v} dt', \qquad\qquad \text{(B.10)}
$$

since $r(x, v, u)$ is the intensity of radiation that is diffusely reflected from the medium, to emerge with direction cosine v. Equation (B.9) gives J at the top in terms of the reflection function. Next, we must obtain the Cauchy problem for r.

B.3 Reflected Intensities

B.3.1 The S function

From our previous experience, we know that it is convenient to introduce the S function, through the relation with r,

$$
r(x, v, u) = \frac{S(x, v, u)}{4v}, \qquad\qquad \text{(B.11)}
$$

or,

$$
S(x, v, u) = 4 \int_0^x J(t', x, u) e^{-(x-t')/v} dt'. \qquad\qquad \text{(B.12)}
$$

We can also express the source function at the top in terms of the S function,

$$
J(x, x, u) = \frac{1}{4}\lambda(x) \left\{ 1 + \frac{1}{2} \int_0^1 S(x, v, u) \frac{dv}{v} \right\}. \qquad\qquad \text{(B.13)}
$$

B.3.2 Imbedding equations

By differentiating Eq. (B.12), we obtain

$$S_x(x, v, u) = 4J(x, x, u) + 4\int_0^x J_x(t', x, u)e^{-(x-t')/v}dt'$$
$$+4\int_0^x J(t', x, u)\frac{-1}{v}e^{-(x-t')/v}dt'.$$

Substitution using differential-integral Eq. (B.8) for the derivative of J and using the preceding equations leads us to the differential-integral equation for the S function:

$$S_x(x, v, u) = 4J(x, x, u) - \frac{1}{u}S(x, v, u) - \frac{1}{v}S(x, v, u)$$
$$+J(x, x, u)2\int_0^1 S(x, v', u)\frac{dv'}{v'}$$

or

$$S_x(x, v, u) = -\left[\frac{1}{u} + \frac{1}{v}\right]S(x, v, u) \tag{B.14}$$
$$+\lambda(x)\left\{1 + \frac{1}{2}\int_0^1 S(x, v', u)\frac{dv'}{v'}\right\}\left\{1 + \frac{1}{2}\int_0^1 S(x, v, u')\frac{du'}{u'}\right\}.$$

This is the differential-integral equation for S which we obtained in Chapter 2 using the physical imbedding method. The initial condition arises readily by setting $x = 0$ in Eq. (B.12),

$$S(0, v, u) = 0. \tag{B.15}$$

B.4 Internal Intensities

B.4.1 Introduction

Internal intensity functions are defined as integrals of the source function. For upwelling directions, we have

$$I(t, v, x, u) = \frac{1}{v}\int_0^t J(t', x, u)e^{-(t-t')/v}dt', \quad 0 \le t \le x; 0 \le v \le 1. \tag{B.16}$$

For downwelling directions, the formula is

$$I(t, -v, x, u) = \frac{1}{v}\int_t^x J(t', x, u)e^{-(t'-t)/v}dt', \quad 0 \le t \le x; 0 \le v \le 1. \tag{B.17}$$

B.4.2 Imbedding equations

To analytically derive the imbedding equations for the upwelling intensity, we differentiate on both sides of Eq. (B.16), obtaining

$$I_x(t, v, x, u) = \frac{1}{v} \int_0^t J_x(t', x, u) e^{-(t-t')/v} dt'. \tag{B.18}$$

This can be written, by using (B.8) and (B.16), in a form that we recognize from Chapter 2, i.e.,

$$I_x(t, v, x, u) = -\frac{1}{u} I(t, v, , x, u) + 2J(x, x, u) \int_0^1 I(t, v, x, u') \frac{du'}{u'}. \tag{B.19}$$

This is one of the desired results. For the initial condition, we set $t = x$ in the definition of Eq. (B.16) and use the S function to obtain the expected equation,

$$I(x, v, x, u) = \frac{S(x, v, u)}{4v}. \tag{B.20}$$

For the downwelling intensity, we differentiate Eq. (B.17), obtaining

$$I_x(t, -v, x, u) = \frac{1}{v} J(x, x, u) e^{-(x-t)/v} \tag{B.21}$$

$$+ \frac{1}{v} \int_t^x J_x(t, x, u) e^{-(t'-t)/v} dt'$$

and

$$I_x(t, -v, x, u) = -\frac{1}{u} I(t, -v, x, u) + 2J(x, x, u) \tag{B.22}$$

$$\cdot \int_0^1 I(t, v, x, u') \frac{du'}{u'} + J(x, x, u) e^{-(x-t)/v},$$

which agrees with that obtained in Chapter 2. The appropriate initial condition is obtained by setting $t = x$ in Eq. (B.17), i.e.,

$$I(x, -v, x, u) = 0. \tag{B.23}$$

B.4.3 Discussion

The equations derived analytically for the internal intensity functions agree exactly with those obtained in Chapter 2 using the physical form of the invariant imbedding method. The internal intensity functions are known to satisfy the equation of transfer (see Chapter 1 and Appendix A), which is a linear two-point boundary value problem. Thus we have shown that the boundary value problem and the Cauchy problem for the internal intensity function are equivalent.

B.5 The Fredholm Resolvent

B.5.1 Resolvent for Fredholm integral equations

The resolvent $K(t, z)$ of the Fredholm integral equation,

$$y(t) = f(t) + \int_0^x k(t, t') y(t') dt' \tag{B.24}$$

is the solution of the integral equation with the forcing function being the kernel itself:

$$K(t, z) = k(t, z) + \int_0^x k(t, t') K(t', z) dt'. \tag{B.25}$$

The resolvent is a useful function to know because the solution with a general forcing function $f(t)$ can be expressed through the relation,

$$y(t) = f(t) + \int_0^x K(t, t') f(t') dt'. \tag{B.26}$$

The resolvent itself can be expressed as

$$K(t, z) = k(t, z) + \int_0^x K(t, t') k(t', z) dt'. \tag{B.27}$$

This implies that

$$\int_0^x k(t, t') K(t', z) dt' = \int_0^x K(t, t') k(t', z) dt'. \tag{B.28}$$

If the original kernel is symmetric, then so is the resolvent.

B.5.2 Resolvent for radiative transfer in a homogeneous slab

In the treatment of radiative transfer, it is customary to indicate the optical thickness x of the medium as a parameter or independent variable. Consider the homogeneous slab with constant albedo for single scattering λ. From the Milne equation, the Fredholm integral equation of radiative transfer with uniform parallel radiation at the top with direction cosine u, we have the symmetric kernel,

$$k(t, t') = \frac{\lambda}{2} E_1(|t - t'|) \tag{B.29}$$

and we have the integral equation for the resolvent,

$$K(t, z, x) = \frac{\lambda}{2} E_1(|t - z|) + \int_0^x \frac{\lambda}{2} E_1(|t - t'|) K(t', z, x) dt'. \tag{B.30}$$

The resolvent function has a physical meaning:

$$K(t, z, x) \quad = \quad \text{the rate of production of scattered radiation,} \quad \text{(B.31)}$$

per unit volume per unit solid angle per unit time per unit frequency interval, at altitude t in a slab of thickness x due to isotropic sources at z.

To derive the integral equation for K, we consider the quantity $K(t, z, x)dt$. The contribution of the radiation directly traveling from z to t is

$$\int_0^1 \frac{1}{4\pi} \exp(|t - z|/v)(dt/v)2\pi dv\lambda = \frac{\lambda}{2} E_1(|t - z|)dt.$$

The indirect contribution of radiation produced at z ultimately being scattered at z' and followed by directly going to t is:

$$\int_0^x K(z', z, x)dz' \frac{\lambda}{2} E_1(|t - z'|)dt = \frac{\lambda}{2} \int_0^x E_1(|t - z'|)K(z', z, x)dz'dt.$$

Putting the three expressions together, we have

$$K(t, z, x)dt = \frac{\lambda}{2} E_1(|t - z|)dt + \frac{\lambda}{2} \int_0^x E_1(|t - z'|)K(z', z, x)dz'dt. \quad \text{(B.32)}$$

Dividing through by dt results in the integral equation for the resolvent,

$$K(t, z, x) = \frac{\lambda}{2} E_1(|t - z|)dt + \frac{\lambda}{2} \int_0^x E_1(|t - z'|)K(z', z, x)dz'. \quad \text{(B.33)}$$

Recall that the Φ function of radiative transfer satisfies the integral equation,

$$\Phi(t, x) = \frac{\lambda}{2} E_1(|t - z|)dt + \frac{\lambda}{2} \int_0^x E_1(|t - z'|)\Phi(z', x)dz'. \quad \text{(B.34)}$$

This means that the resolvent is equal to Φ when one of its arguments is x, i.e., the resolvent for the rate of production at the top of the slab is

$$K(t, x, x) = \Phi(t, x). \quad \text{(B.35)}$$

The resolvent has another physical meaning in probabilistic radiative transfer,

$$K(t, z, x)dt \quad = \quad \text{the probability that a photon absorbed at } z \text{ (B.36)}$$

will be absorbed between t and $t + dt$.

The equivalence of these two interpretations of K is due to the interchangeability of directions in the radiative transfer process. The probabilistic method for radiative transfer has been extensively developed by Sobolev and Ueno [4]–[10].

B.5.3 Invariant imbedding for the resolvent

Let us derive a differential equation for K from its integral equation. We differentiate with respect to x on both sides of the integral equation, and we obtain

$$K_x(t, z, x) = \frac{\lambda}{2} E_1(|t - z|)dt + \frac{\lambda}{2} \int_0^x K_x(t, z', x) E_1(|t - z'|)dz'.$$

Solving this integral equation for the function K_x, we have

$$K_x(t, z, x) = K(t, z, x)K(x, z, x).$$

This equation is known as the Bellman-Krein equation [8, 12]. For an initial condition, we may use

$$K(t, x, x) = \Phi(t, x),$$

where $\Phi(t, x)$ is known through its Cauchy problem involving the source function J. For numerical results obtained by the invariant imbedding method, see Reference [11].

B.6 Discussion

The invariant imbedding Cauchy equations for the source function have been derived analytically starting from the Fredholm integral equation. These equations are exactly the same as those obtained using the physical version of invariant imbedding. Thus these two versions of invariant imbedding are equivalent, and the equations obtained by one are validated by the other.

The imbedding equations for the reflected intensities and the internal intensities have also been derived analytically. They agree completely with those obtained physically.

We have shown that the integral equation, the two-point boundary value problem, and the Cauchy problems of radiative transfer are equivalent. Thus the investigator has the choice of mathematical formulation as well as of computational method. We have, in this book, emphasized the Cauchy formulation that leads to the most effective computational results.

The elegant analytical methods of the preceding sections apply to other cases treated in this book, for example, when there are emitting sources in the medium. See also the treatment of the greenhouse effect in Appendix C.

References

1. R. Courant and D. Hilbert, *Methods of Mathematical Physics*, Interscience Publishers, New York, 1953.

2. H. Kagiwada and R. Kalaba, *Integral Equations via Imbedding Methods*, Addison-Wesley Publishing Co., Reading, Mass., 1974.

3. M. Abramowitz and I. Stegun (eds.), *Handbook of Mathematical Functions*, National Bureau of Standards, 1964.

4. V. Sobolev, "The Transmission of Radiation in an Inhomogeneous Medium," *Doklady Akad. Nauk, SSSR*, Vol. 111, 1956, pp. 1000–1003.

5. S. Ueno, "The Probabilistic Method of Problems of Radiative Transfer: X. Diffuse Reflection and Transmission in a Finite Inhomogeneous Atmosphere," *Astrophys. J.*, Vol. 132, 1960, pp. 729–745.

6. S. Ueno, "The Invariant Imbedding Method for Transport Problems – II. Resolvent in Photon Diffusion Equation," *J. Math. Anal. Appl.*, Vol. 2, 1961, pp. 361–373.

7. V. Sobolev, *A Treatise on Radiative Transfer*, D. Van Nostrand Co., Inc., New York, 1963.

8. R. Bellman, R. Kalaba and S. Ueno, "Invariant Imbedding and a Resolvent of the Photon Equation," *The Rand Corporation*, RM-3937-ARPA, March 1964.

9. R. Bellman, H. Kagiwada, R. Kalaba and S. Ueno, "A New Derivation of the Integro-Differential Equations for Chandrasekhar's X and Y Functions," *J. Math. Phys.*, Vol. 9, 1968, pp. 906–908.

10. R. Bellman, H. Kagiwada, R. Kalaba and S. Ueno, "Analytical and Computational Techniques for Multiple Scattering in Inhomogeneous Finite Slabs," *The Rand Corporation*, RM-4438-PR, January 1965.

11. J. Buell, H. Kagiwada, R. Kalaba, A. McNabb and A. Schumitzky, "Computation of the Resolvent for the Auxiliary Equation of Radiative Transfer," *J. Quant. Spectrosc. Rad. Transfer*, Vol. 8, 1968, pp. 1481–1489.

12. M. G. Krein, "A New Method for Solving Linear Integral Equations of the First and Second Kinds," *Doklady AN SSSR*, Vol. 100, 1955, pp. 413–416.

C. Appendix C
Greenhouse Effect

The greenhouse effect contributes to the warming of the earth's atmosphere. A simple yet productive model has been presented by Milne [1] and Wildt [2] for a gray atmosphere of infinite optical thickness. With the use of invariant imbedding, we determine temperature profiles in model atmospheres of various thicknesses and greenhouse coefficients.

C.1 Introduction

Solar radiation of net flux π is incident uniformly on the top of a plane parallel atmosphere, with incident angle being arc cosine v with respect to the normal to the surface. The model of the greenhouse effect assumes a homogeneous absorbing and scattering atmosphere of finite optical thickness c in which the solar radiation has an absorption coefficient K_s and the diffuse planetary radiation the absorption coefficient K_p. The greenhouse parameter, n,

$$n = K_s/K_p \leq 1, \tag{C.1}$$

is assumed constant. Clearly, when the absorption coefficients for the solar and planetary radiations are the same, then $n = 1$ and there is no greenhouse effect.

We introduce the optical depth τ, where

$$d\tau = -K_p dy, \tag{C.2}$$

and y is the geometrical distance from the bottom. Local thermodynamic equilibrium [3] is assumed, and the local rate of emission of heat is $4\pi K_p B(\tau, v)$, where

$$B(\tau, v) = \frac{Q}{\pi} T^4, \tag{C.3}$$

according to the Stefan-Boltzmann law. The source function $B(\tau, v)$ satisfies the integral equation,

$$B(\tau, v) = \frac{n}{4} e^{-\tau/v} + \frac{1}{2} \int_0^c E_1(|t - \tau|) B(t, v) dt, \tag{C.4}$$

where the kernel involves the first integral function,

$$E_1(r) = \int_0^1 e^{-r/v}\, dv/v. \tag{C.5}$$

The forcing term,

$$g(\tau) = \frac{n}{4} e^{-\tau/v} \tag{C.6}$$

is the reemission of the absorbed solar radiation, while the integral term is the reemission at depth τ of energy previously emitted at depth t, where t varies over all depths.

C.2 Integral Equation for Source Function

We introduce $u(t, x)$, the source function when $g(t)$ expresses the radiation emitted at optical altitude t (rather than depth τ). It satisfies the integral equation,

$$u(t, x) = g(t) + \frac{1}{2} \int_0^x E_1(|t - y|) u(y, x)\, dy, \tag{C.7}$$
$$0 \le t \le x; \quad 0 \le x \le c.$$

In the derivation which follows, the emission function is at first kept quite general. Then, in computing the solution for the greenhouse problem, it is set equal to

$$g(t) = \frac{n}{4} e^{-(c-t)/v}, \tag{C.8}$$

when c is the optical thickness of the slab of interest.

C.3 Invariant Imbedding

We apply the analytical invariant imbedding derivation discussed in Appendix B, in which we vary the thickness by adding a differential layer to the top of the medium. Differentiating with respect to the variable thickness x on both sides of the integral equation for $u(t, x)$, while considering t to be constant, we obtain the integral equation for the function $u_x(t, x)$ with the forcing function

$$\frac{1}{2} E_1(|t - x|) u(x, x),$$

where $u(x, x)$ is the source function at the top of the atmosphere. The Sobolev Φ function is again introduced as the solution of an auxiliary integral equation with forcing function

$$\frac{1}{2} E_1(|t - x|).$$

Then the solution $u_x(t, x)$ is expressed in terms of $\Phi(t, x)$ as

$$u_x(t, x) = u(x, x)\Phi(t, x), \quad 0 \le x \le c. \tag{C.9}$$

This is one of the desired imbedding equations. As we saw in Appendix B, the Φ function is expressible in terms of the source function with parallel rays of illumination, $J(t, x, v)$, namely,

$$\Phi(t, x) = 2 \int_0^1 J(t, x, v) dv/v. \tag{C.10}$$

This function $J(t, x)$ is a particularization to the case of conservative scattering in which the albedo for single scattering is identically unity. The imbedding equations which J satisfies have a special form in terms of two functions, commonly called the Chandrasekhar X and Y functions [4].

The imbedding equation for $J(t, x)$ is

$$J_x(t, x, v) = -\frac{1}{v} + J(x, x, v)\Phi(t, x), \quad x \geq t. \tag{C.11}$$

Anticipating our need for J at the bottom, we set $t = 0$ in the above equation, and we have another imbedding equation to include in the complete Cauchy problem,

$$J_x(0, x, v) = -\frac{1}{v}J(0, x, v) + J(x, x, v)\Phi(0, x), \quad x \geq 0. \tag{C.12}$$

Since this equation contains J at the top, we need to obtain an equation for it. This time we will keep the point of interest at a fixed depth while we vary the thickness by adding a layer to the bottom of the slab. We use the integral equation for J at a fixed depth,

$$J(x - \tau, x, v) = \frac{1}{4}e^{-\tau/v} + \frac{1}{2}\int_0^x E_1(|t - y|)J(x - y, x, v)dy. \tag{C.13}$$

Differentiation leads to the integral equation for $(d/dx)J(x - \tau, x, v)$ which has the forcing function,

$$\frac{1}{2}E_1(|x - \tau|)J(0, x, v).$$

From the superposition principle for solutions of this linear integral equation results the following imbedding equation,

$$\frac{d}{dx}J(x - \tau, x, v) = J(0, x, v)\Phi(\tau, x). \tag{C.14}$$

We set the depth equal to zero and obtain the equation for J at the top,

$$\frac{d}{dx}J(x, x, v) = J(0, x, v)\Phi(0, x), \quad x \geq 0. \tag{C.15}$$

This is another of the desired basic differential equations. At this point we have the basic set of differential equations for computing J at a fixed altitude t, at the bottom, and at the top. We also have one of the equations for the source function at altitude t, in which the source function at the top appears. We next derive the equation for the latter.

The expression we use for $u(x, x)$ is

$$u(x, x) = g(x) + \frac{1}{2} \int_0^x E_1(x - y) u(y, x) dy. \tag{C.16}$$

Using the expression for the E_1 function, we relate the function $u(x, x)$ to $e(x, v)$, the emergent energy per unit horizontal area per unit solid angle per unit time,

$$u(x, x) = g(x) + \frac{1}{2} \int_0^1 e(x, v) dv / v, \tag{C.17}$$

where

$$e(x, v) = \int_0^x e^{-(x-y)/v} u(y, x) dy. \tag{C.18}$$

Differentiation of the left and right sides of eq. (C.18) leads to the equation,

$$
\begin{aligned}
e_x(x, v) &= -u(x, x) - \frac{1}{v} \int_0^x e^{-(x-y)/v} u(y, x) dy + \int_0^x e^{-(x-y)/v} u_x(y, x) dy \\
&= -\frac{1}{v} e(x, v) + u(x, x) \left\{ 1 + \int_0^x e^{-(x-y)/v} \Phi(y, x) dy \right\}.
\end{aligned}
$$

The expression in curly brackets can be evaluated as follows. The integral equation satisfied by the source function $J(t, x, u)$ is given by eq. (B.1). Express the solution of this integral equation using the resolvent formula of (B.26),

$$J(t, x, u) = \frac{1}{4} e^{-(x-t)/u} + \int_0^x K(t, y, x) \frac{1}{4} e^{-(x-y)/u} dy,$$

and evaluate the left and right sides at $t = x$. By (B.35), we know that the $\Phi(y, x)$ function is the resolvent with the first and third arguments the same, namely, $K(x, y, x)$. Then the expression on the right is the expression in curly brackets, and we can replace it by $4J(x, x, v)$.

These steps lead us to the imbedding equation.

$$e_x(x, v) = -\frac{1}{v} e(x, v) + 4u(x, x) J(x, x, v). \tag{C.19}$$

The X and Y functions are introduced as

$$X(x, v) = 4J(x, x, v), \tag{C.20}$$

$$Y(x, v) = 4J(0, x, v). \tag{C.21}$$

To ease the notation, we also introduce the function of one argument for u at the top,

$$A(x) = u(x, x) = g(x) + \frac{1}{2} \int_0^1 e(x, v) dv / v. \tag{C.22}$$

The basic Cauchy problem consists of the coupled equations for X, Y and e which are functions only of thickness and one angle,

$$X_x(x, v) = Y(x, v) \frac{1}{2} \int_0^1 Y(x, v') dv' / v', \tag{C.23}$$

$$Y_x(x, v) = -\frac{1}{v} Y(x, v) + X(x, v) \frac{1}{2} \int_0^1 Y(x, v') dv' / v', \tag{C.24}$$

$$e_x(x, v) = -\frac{1}{v} e(x, v) + A(x) X(x, v). \tag{C.25}$$

To the above we adjoin the additional equations for determining the source functions $u(t, x)$ and $J(t, x, v)$ at altitude t,

$$J_x(t, x, v) = -\frac{1}{v} J(t, x, v) + \frac{1}{2} X(x, v) \int_0^1 J(t, x, v') dv' / v', \tag{C.26}$$

$$u_x(t, x) = A(x) 2 \int_0^1 J(t, x, v') dv' / v'. \tag{C.27}$$

The initial conditions at $x = 0$ are

$$X(0, v) = 1, \tag{C.28}$$

$$Y(0, v) = 1, \tag{C.29}$$

$$e(0, v) = 0. \tag{C.30}$$

The initial conditions when the differential equations are adjoined at altitude t are:

$$[J(t, x, v)]_{x=t} = \frac{1}{4} X(t, v), \tag{C.31}$$

$$[u(t, x)]_{x=t} = g(t) + \frac{1}{2} \int_0^1 e(t, v') dv' / v'. \tag{C.32}$$

C.4 Computational Method

The procedure for determining the source function $u(t, c)$ at a fixed altitude in a slab of finite thickness c consists of the following. First, it is understood that the definite integrals in the Cauchy system are approximated by a Gaussian quadrature formula of order N, where N is an input value. This results in a system of ordinary differential equations for the functions of interest, together with their initial conditions. We solve the $3N$ eqs. (C.23)–(C.25) for X, Y, and e, using eq. (C.22) and the initial conditions in eqs. (C.28)–(C.30) for $0 \leq x \leq t$. Then, when $x = t$, the conditions in eqs. (C.31) and (C.32) are imposed and the $N + 1$ eqs. (C.26) and (C.27) are adjoined for $J(t, x, v)$ and $u(t, x)$. The solution is found for increasing thickness until x finally attains the maximum desired thickness. This is done for any choice of altitude t, $0 \leq t \leq$ the maximum thickness. If there are M altitudes at which the source function is desired, then M times as many equations and initial conditions are involved for the source functions.

C.5 Computational Results

We set the forcing function $g(t)$ equal to that appropriate for the greenhouse model (eq. (C.8)) with an input value of the greenhouse parameter n. We use $N = 7$ and an integration step size of 0.005, together with a fourth order numerical integration method. Some of the results obtained are presented in Figures C.1 and C.2 for slab thicknesses 5, 10, ..., 50 and at selected depths of 0, 5, 10, and 20. Observe the increasing temperature as the thickness increases. The temperatures at different depths reach limiting values. These values agree with those for infinite thickness provided by Professor Wildt to the authors; Wildt's values were obtained from the H functions for infinite thickness.

Clearly, the moderate greenhouse case with $n = 1/10$ displays elevated internal temperature profiles, particularly near the top and especially for thinner atmospheres. This dependence on thickness could not be seen from models that assumed infinite thickness. The comparison with Wildt is shown in ref. [5].

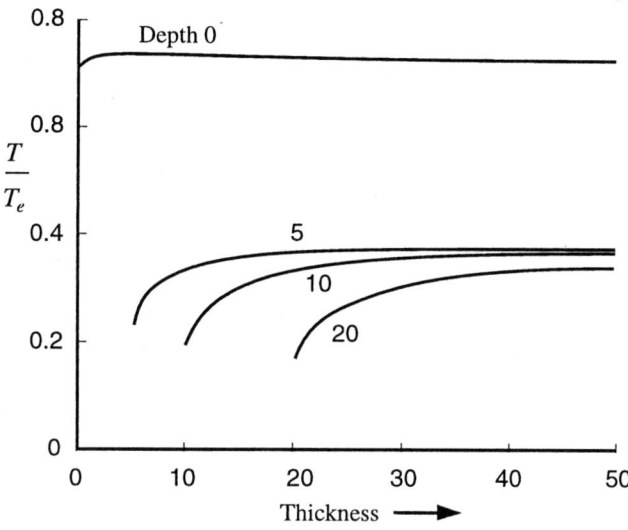

Figure C-1. Temperature at various optical depths as a function of thickness of atmosphere. No greenhouse effect ($n = 1$), incident direction cosine 0.05.

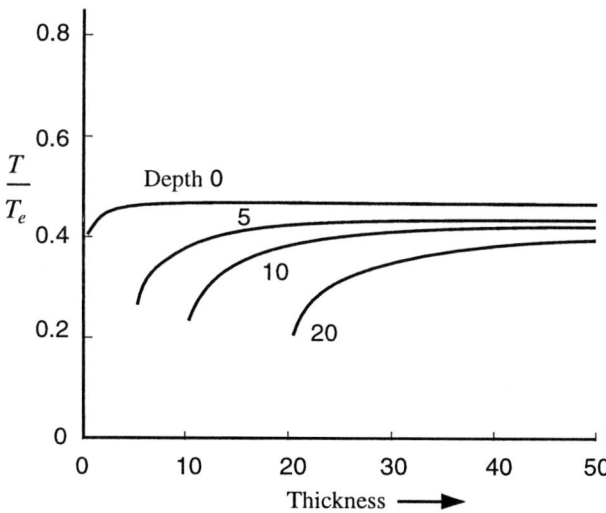

Figure C-2. Temperature at various optical depths as a function of thickness of atmosphere. Moderate greenhouse effect ($n = 1/10$), incident direction cosine 0.05.

References

1. E. A. Milne, *Phil. Mag.*, Vol. 44, 1922, pp. 872–896.
2. R. Wildt, *The Greenhouse Effect in a Gray Planetary Atmosphere*, Icarus, Vol. 5, 1966, pp. 24–33.
3. E. Hopf, *Mathematical Problems of Radiative Equilibrium*, Cambridge Tracts, No. 31, 1934.
4. S. Chandrasekhar, *Radiative Transfer*, Dover Publications, New York, 1960.
5. H. Kagiwada and R. Kalaba, *A New Approach to the Auxiliary Equation with Application to the Greenhouse Effect*, J. Math. and Phys. Sci., Vol. 111, 1969, 176–184.

D. Appendix D
Identification of an Atmospheric Medium

In this appendix we establish a new method to identify the physical properties
of an atmospheric medium, if the input and the output radiations are known
at the boundaries. We consider the atmospheric medium to be plane parallel
and inhomogeneous with anisotropic scattering.

D.1 Introduction

We consider the inverse problem of estimating the properties of a scattering atmosphere based on known or measured data about the incident and
emergent radiation fields. The new method developed here is based on the
properties of the scattering matrix of Chapter 6. To solve the inverse problem
of radiative transfer, we compare two closely related media, one with known
medium coefficients given by matrix $M(z)$ and the other with the matrix of
unknown medium coefficients matrix $\bar{M}(z)$, i.e.,

$$M(z) = \begin{pmatrix} b & a \\ c & d \end{pmatrix} \quad \text{and} \quad \bar{M}(z) = \begin{pmatrix} \bar{b} & \bar{a} \\ \bar{c} & \bar{d} \end{pmatrix} \tag{D.1}$$

where $0 \leq z \leq h$ and h is the optical thickness of both media; see [2].
We take advantage of the fact that scattering matrices of radiative transfer
are dissipative [1], [2]. We are able to establish relations of these scattering
matrices associated with two medium coefficients. This relationship provides
our solution to the identification problem. For a discussion of related topics
and methods, see [3]–[9] and chapters 3, 6, 7 and 8.

We present here a step-by-step answer to solve this problem.

D.2 Scattering Matrix from Matrix of Medium Coefficients $M(z)$

We wish to determine the scattering matrix $\mathfrak{S}(0, h)$ from $M(z)$. In theory,
$\mathfrak{S}(0, h)$ can be obtained directly by solving the multiple scattering problem
via an invariant imbedding method discussed in Chapter 6. However, we
present an approximation method here. We partition the atmosphere into

n layers so that their thicknesses are Δz_i. Let us set $z_{i+1} = z_i + \Delta z_i$ and $\mathfrak{S}(z_i, z_{i+1}) = \mathfrak{S}_i$, $i = 1, 2, \ldots, n$, where $\mathfrak{S}_i$ is the scattering matrix extended from $z = z_i$ to $z = z_{i+1}$. For Δz_i small then, we have

$$\mathfrak{S}_i = E + M(\bar{z}_i)\Delta z_i = E + \begin{pmatrix} b(\bar{z}_i) & a(\bar{z}_i) \\ c(\bar{z}_i) & d(\bar{z}_i) \end{pmatrix} \Delta z_i, \qquad (D.2)$$

where E is the identity matrix and $z_i \leq \bar{z}_i \leq z_{i+1}$. Since, see Chapter 6,

$$\mathfrak{S}(0, h) = \mathfrak{S}_1 * \mathfrak{S}_2 \cdots * \mathfrak{S}_n$$

we obtain

$$\mathfrak{S}(0, h) = [E + M(\bar{z}_1)] * [E + M(\bar{z}_2)] * \cdots * [E + M(\bar{z}_n)], \qquad (D.3)$$

for an approximation, letting $\bar{z}_i = (z_i + z_{i+1})/2$. As $n \to \infty$, we obtain exact $\mathfrak{S}(0, h)$.

Eq. (D.3) requires many $*$-product multiplications. Let us define the hat operation

$$\hat{\mathfrak{S}} = -\begin{pmatrix} \tau & 0 \\ r & E \end{pmatrix}^{-1} \begin{pmatrix} -E & \rho \\ 0 & \tau \end{pmatrix},$$

where t, τ, ρ and r are sub-matrices of

$$\mathfrak{S} = \begin{pmatrix} t & \rho \\ r & \tau \end{pmatrix},$$

see chapter 6 for details. The transmissions t and τ, reflections ρ and r are linear operators. It follows that

$$\mathfrak{S}(0, h) = (\hat{\mathfrak{S}}_1 \cdot \hat{\mathfrak{S}}_2 \cdots \hat{\mathfrak{S}}_n)^{\wedge} \qquad (D.4)$$

where $\cdot$ is the usual matrix multiplication. For Δz_i small,

$$\hat{\mathfrak{S}}_i = [E + M(z_i)\Delta z_i]^{\wedge} + o(\Delta z_i). \qquad (D.5)$$

Eq. (D.5) can be reduced to

$$\hat{\mathfrak{S}}_i = \begin{pmatrix} E & 0 \\ 0 & E \end{pmatrix} + \begin{pmatrix} -b_i & -a_i \\ c_i & d_i \end{pmatrix} \Delta z_i + o(\Delta z_i) \qquad (D.6)$$

where $o(\Delta z_i)/\Delta z_i \to 0$ as $\Delta z_i \to 0$.

Eq. (D.5) requires $(E + b_i \Delta z_i)^{-1}$ to exist. Since b_i is a bounded operator, this condition is always satisfied for Δz_i small. The right-hand side of eq. (D.4) is much easier to compute than that of eq. (D.3). The conclusion is, given $M(z)$ for all $0 \leq z \leq h$, we can obtain or estimate $\mathfrak{S}(0, h)$ from (D.3) or (D.4).

D.3 Scattering Matrix from Inputs and Outputs

Let $\bar{\mathfrak{S}}(0, h)$ be the scattering matrix associated with the incident and emergent radiation fields,

$$\tilde{\mathfrak{S}}(0, h) = \begin{pmatrix} \tilde{t} & \tilde{\rho} \\ \tilde{r} & \tilde{\tau} \end{pmatrix} \tag{D.7}$$

and

$$\bar{\mathfrak{S}}(0, h) \cdot \begin{pmatrix} I^+(0) \\ I^-(h) \end{pmatrix} = \begin{pmatrix} I^+(h) \\ I^-(0) \end{pmatrix} \tag{D.8}$$

where $I^{\pm}(0)$ and $I^{\pm}(h)$ are physically measured, i.e., the incident and emergent radiation fields, at boundaries where $\tilde{t}$, $\tilde{\tau}$ are transmission operators and $\bar{\rho}$, $\bar{r}$ are reflection operators associated with an unknown medium.

For our problem operators $\tilde{t}$, $\tilde{\tau}$, $\bar{\rho}$ and $\bar{r}$ are linear operators, for example

$$\begin{aligned} \tilde{t}I^+(h, \Omega') &= \int \tilde{t}(\Omega - \Omega')I^+(h, \Omega')d\Omega' \tag{D.9} \\ &= I^+(0, \Omega) \end{aligned}$$

where $\bar{t}(\Omega - \Omega')$ is the kernel of $\bar{t}$, $\Omega = (\theta, \varphi)$ and $I^+(h, \Omega')$ is the input and $I^+(0, \Omega)$ is the transmitted output. Knowing $I^+(h, \Omega')$ and $I^+(0, \Omega)$, the transmission operator, $\bar{t}$, can be found by convolution, i.e.,

$$\bar{t} = \mathcal{F}^{-1}[\mathcal{F}(\bar{t})] = \mathcal{F}^{-1}\left\{ \frac{1}{2\pi} \mathcal{F}[I^+(0, \Omega)]/\mathcal{F}[I^+(h, +\Omega)] \right\} \tag{D.10}$$

where $\mathcal{F}$ is the Fourier transform. In practice, we may estimate $\tilde{t}$ by using the fast-Fourier series, or by taking discrete values of inputs and outputs, and likewise for other operators. In conclusion, by measurements of inputs and outputs at the boundaries, we obtain $\tilde{\mathfrak{S}}(0, h)$ of (D.7).

D.4 Relationship between Scattering Matrices

If the scattering matrix is dissipative, t and τ are nonsingular, by Theorem 6 of [1], there are two unitary matrices $U_1(z)$ and $U_2(z)$ such that

$$U_1(z) * \mathfrak{S}(0, z) * U_2(z) = D(0, z), \quad 0 \le z \le h \tag{D.11}$$

where D is a diagonalized scattering matrix,

$$D(0, z) = \begin{pmatrix} t_0(0, z) & 0 \\ 0 & \tau_0(0, z) \end{pmatrix}. \tag{D.12}$$

Since $U_1(z)$ and $U_2(z)$ are unitary, it follows that t and τ nonsingular implies t_0 and τ_0 are nonsingular. Also $D(0, z)$ has an inverse under the $*$ product, and

$$U_1(z) * \mathfrak{S}(0, z) * U_2(z) * D^{-1}(0, z) = E. \tag{D.13}$$

Likewise $\bar{\mathfrak{S}}(0, z)$ can be diagonalized to $\bar{D}(0, z)$. And we can solve for $\bar{\mathfrak{S}}(0, z)$ in terms of $\mathfrak{S}(0, z)$, i.e.,

$$\bar{\mathfrak{S}}(0, z) = P(0, z) * \mathfrak{S}(0, z) * Q(0, z) \tag{D.14}$$

where

$$P(0, z) = \bar{U}_1^{-1}(z) * \bar{D}(0, z) * U_1(z) \tag{D.15}$$

and

$$Q(0, z) = U_2(z) * D^{-1}(0, z) * \bar{U}_2^{-1}(z). \tag{D.16}$$

As $x \to 0$, $\bar{\mathfrak{S}}(0, z)$ and $\mathfrak{S}(0, z) \to E$ and

$$P(0, 0) * Q(0, 0) = E. \tag{D.17}$$

Let N and $\bar{N}$ be the infinitesimal generators of $\mathfrak{S}(0, z)$ and $\bar{\mathfrak{S}}(0, z)$, respectively, i.e.,

$$\frac{\partial \hat{\mathfrak{S}}(0, z)}{\partial z} = \hat{\mathfrak{S}}(0, z) N(z). \tag{D.18}$$

and

$$\frac{\partial \hat{\bar{\mathfrak{S}}}(0, z)}{\partial z} = \hat{\bar{\mathfrak{S}}}(0, z) \bar{N}(z). \tag{D.19}$$

for $0 \le z \le h$, see Chapter 6. Taking the $\hat{\,}$ operation in both sides of eq. (D.15) and differentiating with respect to z, we have

$$\begin{aligned} \frac{\partial \hat{\bar{\mathfrak{S}}}(0, z)}{\partial z} &= \hat{P}(0, z) \cdot \frac{\partial \hat{\mathfrak{S}}(0, z)}{\partial z} \cdot \hat{Q}(0, z) \\ &+ \frac{\partial \hat{P}(0, z)}{\partial z} \cdot \hat{\mathfrak{S}}(0, z) \cdot \hat{Q}(0, z) \\ &+ \hat{P}(0, z) \cdot \hat{\mathfrak{S}}(0, z) \cdot \frac{\partial \hat{Q}(0, z)}{\partial z}. \end{aligned} \tag{D.20}$$

On the other hand if $\bar{N}(z)$ satisfies the equation

$$\hat{Q}(0, z) \cdot \bar{N}(z) = N(z) \cdot \hat{Q}(0, z) \tag{D.21}$$

then by (D.21), (D.18) and taking $\hat{\,}$ operation on both sides of eq. (D.14),

$$\begin{aligned} \hat{\bar{\mathfrak{S}}}(0, x) \cdot \bar{N}(x) &= \hat{P}(0, x) \cdot \hat{\mathfrak{S}}(0, x) \cdot \hat{Q}(0, x) \cdot \bar{N}(x) \\ &= \hat{P}(0, x) \cdot \hat{\mathfrak{S}}(0, x) \cdot N(x) \cdot \hat{Q}(0, x) \\ &= \hat{P}(0, x) \cdot \frac{\partial \hat{S}(0, x)}{\partial x} \cdot \hat{Q}(0, x) \end{aligned} \tag{D.22}$$

Comparing equations (D.22), (D.20) and (D.19), it is necessary that the last two terms of equation (D.20) be zeros, i.e.,

$$\frac{\partial \hat{P}(0, z)}{\partial z} \cdot \hat{\mathfrak{S}}(0, z) \cdot \hat{Q}(0, z) + \hat{P}(0, z) \cdot \hat{\mathfrak{S}}(0, z) \cdot \frac{\partial \hat{Q}(0, z)}{\partial z} = 0. \qquad \text{(D.23)}$$

A sufficient condition for $\hat{P}$ and $\hat{Q}$ to satisfy eq. (D.23) is that

$$\frac{\partial \hat{P}(0, z)}{\partial z} = \frac{\partial \hat{Q}(0, z)}{\partial z} = 0. \qquad \text{(D.24)}$$

That is, $\hat{P}$ and $\hat{Q}$ are independent of z. In view of their initial conditions in eq. (D.17), we conclude that

$$\hat{Q}^{-1} = \hat{P} \qquad \text{(D.25)}$$

and

$$\bar{\mathfrak{S}}(0, z) = Q^{-1} * \mathfrak{S}(0, z) * Q \qquad \text{(D.26)}$$

for all $0 \le z \le h$. We say $\bar{\mathfrak{S}}$ and $\mathfrak{S}$ are similar matrices under the $*$-operation. By the analyses presented in D.2 and D.3, $\mathfrak{S}(0, h)$ and $\bar{\mathfrak{S}}(0, h)$ are known. Substituting these values into equation (D.26) at $z = h$, we either estimate or compute the matrix Q. After we obtain Q, equations (D.25) and (D.26) provide us answers for all $\bar{\mathfrak{S}}(0, z)$, $0 \le z \le h$. And, by Chapter 6,

$$\bar{M}(z) = \lim_{\Delta z \to 0} \frac{\bar{\mathfrak{S}}(z + \Delta z) - E}{\Delta z}, \qquad 0 \le z \le h. \qquad \text{(D.27)}$$

Our identification problem is solved.

D.5 Conclusion

We have established an analytical method to determine the medium coefficients of an unknown medium by the invariant imbedding and scattering theorem. We have not developed computational procedures and sample results. When $\bar{M}(z)$ and $M(z)$ are "near" each other, operators $P(z)$ and $Q(z)$ are "near" to the identity. Therefore it is wise to choose $M(z)$ "near" $\bar{M}(z)$. Interested readers should compare results developed here with those in Chapters 3, 6, 7 and 8. Chapters 3 and 4 engage systematic computational methods for a variety of scattering systems. In Chapters 7 and 8, inverse problems were oriented to determine the unknown boundary conditions for remotely sensed data.

References

1.   A. P. Wang, "Dissipative Properties of a Nonlinear Differential Operator Equation," *J. Mathematical Analysis and Appl.*, Vol. 72, No. 1, Nov. 1979, pp. 75–88.

2.   A. P. Wang, "Searchlight on a Target with Diffuse Background," *J. Math. Physics*, Vol. 34, 1993, pp. 878–884.

3.   R. Bellman and S. Ueno, "Invariant Imbedding and Chandrasekhar's Planetary Problem of Radiative Transfer," *Astrophys. Space Sci.*, Vol. 16, 1972, pp. 241–248.

4.   A. Bruckstein and T. Kailath, "An Inverse Scattering Framework for Several Problems in Signal Processing," *IEEE ASSP*, Jan. 1987, pp. 6–20.

5.   R. Redheffer, "On the Relation of Transmission Theory to Scattering and Transfer," *J. Mathematics and Physics*, Vol. XLI, No. 1, March 1962, pp. 1–41.

6.   S. Ueno, R. Kalaba, H. Kagiwada and R. Bellman, "Some Mathematical Aspects of Multiple Scattering in a Finite Inhomogeneous Slab with Anisotropic Scattering," *Pub. Astr. Soc. Japan*, Vol. 22, 1970, pp. 75–83.

7.   S. Ueno and Y. Kawata, "Earth-atmosphere System Identification of Optical Parameters," *Concise Encyclopedia of Environmental System*, P.C. Young, editor, Pergamon Press, Oxford, 1993, pp. 163–168.

8.   S. Ueno and A. P. Wang, "Scattering and Transmission Functions of Radiation by Finite Atmosphere with Reflecting Surface," *J. Astrophysics and Space Science*, Vol. 23, 1973, pp. 205–219.

9.   A. P. Wang, C. Y. Wang and S. Ueno, "Identification of a Dissipative System," to appear.

Index

A

Absorption coefficient 12, 237
Aerosol attenuation profiles 159
Albedo for single scattering 12, 238
Ambarzumian's principle 135
Anisotropic 239
Attenuation coefficient 240
Attenuation of the intensity 244
Associative memories 74
Associative memory neural networks 74–81
Atmospheric correction 155ff
Automatic derivative evaluation 87

B

Bellman-Krein's formula 125
Bellman-Krein equation 262
Bidirectional reflectance 180
Bremmer series 145

C

Cauchy problem 10
Christoffel weights 27
Coherent scattering 239
Conservative Case 238
Convolution method 161
Cross ratio 150
Cumulative scattering 130
Cumulative source function, M 128
Cumulative transmission 130

D

Density of radiation 242
Diagonal scattering matrix 144

Diffusely transmitted radiation 25
Diffusely reflected radiation 21
Discrete time-dependent case 152
Dissipative 140
Direct problems 20
Dynamic programming 96

E

Emission coefficient 239
Equation of transfer 243
Euler's method 5, 28
Euler procedure 6
Emergent intensities 52
Emission coefficient 239

F

FEED 74, 87–90
Finite orders of scattering 119
Fixed-point 141
Flux 110–114
Fredholm integral equation 252
Fredholm resolvent 259

G

Gaussian quadrature formula 27
Generator 142
Geometrical thickness 238
Greenhouse effect 263